Conversion Factors

LENGTH
1 cm = 0.3937 in. = 0.0328 ft = 6.214 × 10⁻⁴ mile
1 in. = 2.54 cm
1 mile = 1609 m
1 Å (angstrom unit) = 10⁻⁸ cm = 10⁻⁴ micron

AREA
1 m² = 10.76 ft² = 1550 in.² = 2.471 × 10⁻⁴ acre

VOLUME
1 m³ = 35.31 ft³ = 1000 liters = 6.102 × 10⁴ in.³

TIME
1 year = 365.24 days = 8.766 × 10³ hr = 5.259 × 10⁵ min
       = 3.156 × 10⁷ sec

VELOCITY
1 m/sec = 3.281 ft/sec = 3.6 km/hr = 2.237 mile/hr
        = 1.944 knots

MASS
1 kg = 2.205 lb = 1000 g

PRESSURE
1 bar = 0.9689 atm = 14.5 lb/in.² = 75.01 cmHg
      = 401.5 in H₂O = 1 × 10⁶ dynes/cm²

ENERGY
1 joule = 0.2389 cal = 9.481 × 10⁻⁴ Btu
        = 0.7376 ft-lb = 10⁷ ergs
1 liter-atm = 24.21 cal
1 cal = 4.184 joules

CHEMICAL EQUILIBRIA IN THE EARTH

Wallace S. Broecker

Professor of Geology
Lamont Doherty Geological Observatory
of Columbia University

Virginia M. Oversby

Research Fellow, Department of Geophysics and Geochemistry
Research School of Physical Sciences
The Australian National University

McGraw-Hill Book Company

New York
St. Louis
San Francisco
Düsseldorf
London
Mexico
Panama
Sydney
Toronto

Chemical Equilibria in the Earth

Library of Congress Catalog Card Number 79-109246

07997

1 2 3 4 5 6 7 8 9 0 M A M M 7 9 8 7 6 5 4 3 2 1 0

This book was set in Modern by The Maple Press Company, and printed on permanent paper and bound by The Maple Press Company. The designer was Merrill Haber; the drawings were done by B. Handelman Associates, Inc. The editors were Bradford Bayne and James W. Bradley. Peter D. Guilmette supervised production.

Preface

Applications of thermodynamic and kinetic arguments to chemical systems in the earth sciences have become so widespread that no graduate student can consider himself equipped for future professional work unless he is able to understand them. This book is intended to be an introduction to the subject within the grasp of the average undergraduate science major.

The first five chapters provide an introduction to thermodynamic logic, both from the macroscopic and atomic points of view. It is our impression that the usual development of the subject is sufficiently abstract to force all but the best student to a recipe approach when attempting to apply his knowledge. It is our hope that our more direct development will help to overcome this difficulty. However, we certainly do not wish to pass off our treatment as a substitute for a standard thermodynamics text. Our coverage is clearly more limited and its development considerably less rigorous. For students who plan to go further into the subject, exposure to a standard thermodynamics text is an absolute necessity. Since thermodynamics is one of those subjects learned only by repetitive exposure, most students will find our introduction to the subject a worthwhile supplement to other training in this area. In any case, these five chapters provide the theory needed to treat most of the applications we deem important.

The remaining seven chapters deal with the most common types of chemical reactions encountered in the earth sciences. In each case at least one common process taking place within our solar system is treated as an example. Both equilibrium and kinetic aspects are considered. Problems at the end of each chapter are designed both to "cement" the understanding of the material covered and to broaden the scope of application to earth phenomena.

Although our book is designed for a one-semester course, the course could easily be expanded to a full year by using the supplementary readings given at the end of each chapter in conjunction with current journal articles. (In the supplementary readings, as well as in the problems sections, items of more than average complexity are marked with an asterisk.) Our intent was to present the bare bones of the subject—not to review the successes and failures of attempts to apply this approach to furthering our knowledge of earth processes and earth history.

We have received the assistance of many people during the preparation of the manuscript, people without whose assistance the work would never have been finished. G. D. Garlick, H. Greenwood, P. W. Gast, and many others made helpful suggestions on how the manuscript could be improved. During classroom trials with earlier versions of the manuscript a number of students spotted ambiguities and errors which have (we hope) been successfully eliminated. H. C. Helgeson and D. R. Waldbaum allowed us to use prepublication data which were vital to the presentations in Chaps. 11 and 12. We are extremely grateful for their generosity. Last, but certainly not least, we thank M. L. Zickle, our long-suffering typist, for the miraculous conversion of illegible notes into neat typescript, and T. Zimmerman for the drafting of the figures.

Wallace S. Broecker
Virginia M. Oversby

Contents

Table of Constants
and Conversion Factors

Avogadro's number	N	6.023×10^{23} mole^{-1}
Boltzmann's constant	k	1.381×10^{-16} erg/deg
Planck's constant	h	6.625×10^{-27} erg-sec
Gas constant	R	1.987 cal/deg-mole
		8.315 joules/deg-mole
		82.06 cm^3-atm/deg-mole
Speed of light	c	2.998×10^{10} cm/sec
Absolute zero		$-273.15°C$
Charge on the electron	e	1.602×10^{-19} coulomb
Mass of the electron		9.109×10^{-28} g
Mass of the proton		1.673×10^{-24} g
Mass of the neutron		1.675×10^{-24} g
Base of natural logarithm	e	2.718
ln 10		2.303

CONVERSION FACTORS

LENGTH

1 cm $= 0.3937$ in. $= 0.0328$ ft $= 6.214 \times 10^{-4}$ mile
1 in. $= 2.54$ cm
1 mile $= 1609$ m
1 Å (angstrom unit) $= 10^{-8}$ cm $= 10^{-4}$ micron

AREA

1 m^2 $= 10.76$ ft^2 $= 1550$ in.^{2} $= 2.471 \times 10^{-4}$ acre

VOLUME

1 m^3 $= 35.31$ ft^3 $= 1000$ liters $= 6.102 \times 10^4$ in.^{3}

TIME

1 year $= 365.24$ days $= 8.766 \times 10^3$ hr $= 5.259 \times 10^5$ min
          $= 3.156 \times 10^7$ sec

VELOCITY

1 m/sec $= 3.281$ ft/sec $= 3.6$ km/hr $= 2.237$ mile/hr
          $= 1.944$ knots

MASS

1 kg $= 2.205$ lb $= 1000$ g

PRESSURE

1 bar $= 0.9689$ atm $= 14.5$ lb/in.^{2} $= 75.01$ cmHg
          $= 401.5$ in H$_2$O $= 1 \times 10^6$ dynes/cm^2

ENERGY

1 joule $= 0.2389$ cal $= 9.481 \times 10^{-4}$ Btu
          $= 0.7376$ ft-lb $= 10^7$ ergs
1 liter-atm $= 24.21$ cal
1 cal $= 4.184$ joules

Chemical Equilibria in the Earth

chapter one Thermodynamics as a Geologic Tool

Research in the geological sciences is directed toward understanding the processes which have contributed to the development of the earth and its fellow planets. Although some of these processes can be effectively observed in action today, others occur so slowly, or in such remote places, that effective observation is not possible. Still other processes took place only in early periods of planetary history. To gain an understanding of these unobservable phenomena, geologists have concentrated their attention on the products of the processes. The fossil remains and detrital mineral grains in sedimentary rocks provide clues regarding paleoenvironment and climate; the textures and minerals of igneous and metamorphic rocks bear record of processes occurring deep within the crust and mantle of the earth; magnetic imprints tell us of slow but persistent global movements of large crustal plates. The problem is to learn to read these records accurately and reliably. Obviously there are innumerable pitfalls in any such endeavor. The more independent approaches that can be brought to bear on a given problem, the greater are the chances that the pitfalls will be overcome and a meaningful solution achieved.

One approach which has captured the fancy of earth scientists studying a wide variety of phenomena is that of chemical thermodynamics. The idea is that, if, at the time a given material formed, its atomic constituents achieved their most stable chemical form and if after formation this form remained unaltered, then the chemical configuration of the elemental and isotopic species present should define the environmental conditions at the time of origin. Temperatures and pressures obtained in this way would, of course, be invaluable in reconstructing earth and planetary histories.

Although perfectly valid in concept, thermodynamic methods suffer from the fact that their two basic assumptions are to some extent mutually exclusive. If at the time of its formation the atoms in a material were capable of intermixing to the extent required to yield the most stable configuration, it is unlikely that, once the material was formed, this mixing would completely cease. The instant quench or "freeze-in" technique employed in laboratory studies rarely occurs in nature. On the other hand, if a substance proves immune to secondary alteration, there is usually some question regarding the equilibration of its constituent atoms with the surroundings at the time of formation. Like other approaches, the thermodynamic one is far from infallible.

We begin our book by introducing the basic thermodynamic parameters needed in stability calculations. By using these parameters it is possible to show how the stable chemical configuration of a system will vary with environmental conditions. Knowing the ideal state of a system formed under a given set of conditions and then perfectly preserved is not enough. In order to evaluate the pitfalls, we must also know something of the rates at which the initial ideal equilibrium would be approached under various environmental conditions and also how rapidly it would be destroyed during storage under some other set of environmental conditions. Thus we look briefly at the very complex subject of kinetics.

Armed with this knowledge, we will consider in detail some of the fundamental processes taking place in the earth and on its surface. In what way do they leave their thermodynamic imprint? What kinetic pitfalls are there? In addition to the examples in the text, problems at the end of each chapter will prove helpful in grasping the principles presented. Selected readings of journal articles provide insight into the current status of attempts to apply these approaches to real geologic problems.

Before we launch out into this course, it will prove helpful to adopt some of the terminology and conventions normally used in the field of thermodynamics. In the discussions which follow, the *system* is defined as that part of the universe which is under primary consideration. The *surroundings* are then defined to be any part of the universe not included in the system. The *state* of the system is specified by listing its properties such as temperature, pressure, chemical composition, and any other convenient parameters. The complexity of the system determines the number of properties necessary to specify completely its state. Any property which depends only on the state of the system, and not on the manner in which that state was achieved, is called a *variable of state*, or *state function*.

The concept of equilibrium, and the limitations which it places on our ability to analyze systems, is of primary importance. The equilibrium state of a system is the state of maximum stability under a given set of conditions. Once a system achieves this state of maximum stability, there is no tendency for it to change its state. Thus a characteristic of the equilibrium state is that its measurable properties do not change with time. However, this does not imply that the system is at rest on the molecular scale. It is true only that any change which takes place in the system on the molecular scale is balanced by changes in the opposite direction, so that we macroscopically observe no change in physical or chemical properties of the system as a whole. This condition defines the state of dynamic equilibrium.

Any nonequilibrium state of a system is unstable with respect to the equilibrium state. This relative instability causes the system to change its properties with time until it reaches the equilibrium state. The rate of this change may be so slow in some cases that we cannot perceive it, but the *tendency* for change exists, nevertheless. On the molecular scale this means that changes in both directions are occurring, but there is a net change in favor of the direction toward equilibrium.

The factors which influence equilibrium depend on the complexity of the system, but in most cases only three factors are of significance. These factors are intimately related to the number of variables which must be specified in order to describe completely the state of a system. The simplest case is that of a pure liquid or gas whose state can be specified completely by giving the value of any two of its variables of state.

The most convenient properties to use are those which can be easily determined experimentally. These include temperature, pressure, and specific volume or its reciprocal, density. Any other property of the system may then, in principle, be calculated from the specified parameters. It is important to realize that, although temperature and pressure, temperature and density, or pressure and density are the common variables of state, the state of the system is just as completely defined by giving its refractive index and viscosity.

In the case of the liquid or gas discussed above, the state which we have defined is the equilibrium state. If the system is not at equilibrium we cannot relate its properties theoretically, and the value of pressure and density will not fix the temperature. In fact, if the system is not in equilibrium there may be temperature differences within the system, in which case there would not be a well-defined temperature of the system as a whole. This represents a major limitation to our ability to analyze systems. We are restricted to discussing systems at equilibrium, since this is the only state for which we can completely describe the system by listing a finite number of parameters. If a system initially at equilibrium is perturbed by a change in conditions and passes through nonequilibrium states to a new equilibrium state determined by these new conditions, we are powerless to describe *how* it achieved its new equilibrium condition. However, we can completely describe this new equilibrium state with a limited amount of information. Fortunately, information concerning the conditions of the final equilibrium state is all that is required in most problems of geologic interest.

There is one special case in which we can describe all intermediate states between an initial and a final equilibrium state. This is when perturbation of the system is gentle enough and gradual enough so that all intermediate states are also equilibrium states. Examples of equilibrium processes are the expansion of a gas against a pressure very slightly less than its own pressure or the melting of a solid at the temperature of its melting point.

We can now consider in more detail the factors which influence the equilibrium state of a system. As can be seen from the preceding discussion, these factors are precisely the same factors which must be specified to fix completely the state of the system. Variables of state may be split into two major subdivisions. Those which depend on the total amount of material in the system are called *extensive variables;* those which are independent of the amount of material are *intensive variables*. Temperature, pressure, and relative abundance of chemical components are examples of intensive variables. Extensive variables, such as volume or heat capacity, may be treated as intensive variables when they are normalized by giving their value relative to a specific amount of material, usually 1 gram (e.g., specific volume) or 1 mole (e.g., molar volume).

Experience has shown that it is usually sufficient to give the chemical composition of a system and the value of two of its intensive variables in order to specify completely the state of the system. The variables usually chosen for geologic problems are temperature and pressure. Since by definition the state of a system must be completely reproducible when enough of its properties are fixed, we can test whether our specification is complete. Let us take a known mass of homogeneous fluid and measure its temperature and pressure. If our theory is correct, this should completely specify the fluid's state and therefore fix the values of all other properties of the fluid.   To check this, let us measure several other properties of the system, for instance, the density, refractive index, and isothermal compressibility.   Now we subject the system to a series of arbitrary perturbations of any magnitude, with the restriction that no material may be added.   On returning the system to its original temperature and pressure, we find that all its physical properties are the same as they were before the series of perturbations.   We can therefore conclude that the state of our system (at constant composition) was completely specified by two of its intensive variables, namely, temperature and pressure.

If any of the physical properties were found to be different at the end of our experiment, we must conclude that our specification of the state is not complete.   This situation arises in the case of solids, where strain history is of importance.   Once our system is completely specified, we can collect the information into a convenient form known as the *equation of state*.   Usually this "equation" consists of tabulated data, but in certain idealized situations it may be represented as an analytic function.

We have seen that intensive or normalized extensive variables are the factors which determine the equilibrium state of a system of constant composition.   Any change in these variables will change the nature of the equilibrium state of our system.   We can greatly simplify our problem by deciding what variations imposed from the surroundings are of major significance in natural situations and how these variations affect the state of our system.

The temperature of the surroundings is certainly of major importance, since the system must eventually adjust itself to be in thermal equilibrium internally and at its boundaries.   The pressure applied on the system by the surroundings is also important, since the system must eventually attain mechanical equilibrium.   The last factor of general importance is the ability of the surroundings to exchange matter with the system, since the properties of the system depend on its composition.

The effects of temperature, pressure, and composition of the surroundings on the system can, of course, be discussed in terms of the temperature, pressure, and composition of the system.   However, as will be seen in Chaps. 2 to 5, the system can be more conveniently described in terms of defined thermodynamic functions which are themselves dependent on temperature, pressure, and composition.   These functions are related to the energy content of the system, its degree of disorder, and its ability to produce useful work.

In the discussions which follow, we will attempt to relate the macroscopic condition of the system, which is the realm of thermodynamics, to the microscopic condition of the system, which is the realm of statistical mechanics and atomic physics. We hope that this integration of methods will give the reader more of an intuitive feeling for systems than would a purist approach of either extreme.

LIMITATIONS OF EQUILIBRIUM CONSIDERATIONS

The thermodynamic study of geologic systems presents several difficulties which must be thoroughly understood before meaningful and realistic conclusions can be drawn. The most obvious problem is that thermodynamics can describe only systems which are at equilibrium or were formed under equilibrium conditions. There is no guarantee that a geologic system was ever totally in equilibrium with its surroundings. We must examine each system carefully to find clues which will indicate whether the system was evolved in a state of internal equilibrium. If this is the case, it is likely that the system was also near equilibrium with its surroundings.

The final judgment that a system evolved under equilibrium conditions involves the use of physical intuition and, in a sense, knowing the answer before we ask the question. However, the situation is not really as bad as it first seems. By examining a fair cross section of geologic conditions for consistency with theoretical predictions, we can gain insight into the nature of geologic systems under equilibrium or near-equilibrium conditions. No process takes place under total equilibrium conditions, since there is no tendency to depart from the equilibrium state. What we need to know is the sensitivity of our results with respect to large departures from equilibrium.

One of the major reasons for large departures from equilibrium conditions is sluggishness of response of the system to changes in its environment. The rates of chemical reactions become, in most cases, the controlling factor in whether systems come to equilibrium. Hydrogen and oxygen gases are thermodynamically unstable with respect to formation of water. However, in the absence of a catalyst or spark, H_2 and O_2 are not observed to react. It *cannot* be concluded that this system is stable; it is merely not labial. That is, the rate of reaction of H_2 and O_2 to give H_2O is so slow in the absence of a catalyst that we cannot perceive its taking place. We must consider experimental data concerning the rates of chemical reactions in the system under examination before we can draw any conclusions about whether the system was formed under equilibrium conditions.

As will be discussed in Chap. 5, chemical reactions become much more rapid at elevated temperatures. Thus the chances for our system to attain internal equilibrium become greater as the temperature of the system becomes greater.   Although it seems a disadvantage at first that chemical reactions are slow at room temperature, it is this very fact that makes useful the study of mineral assemblages as chemical systems. When we examine a rock which was formed at elevated temperatures we see a "frozen-in" record of the conditions under which the rock was formed.   As a rock cools, the rate of chemical reactions to form more stable configurations drops very sharply.   Thus we now observe phases in rocks which are highly unstable at room temperature and atmospheric pressure but which persist because their rate of reaction to form more stable phases is infinitesimally small.

Some reactions are extremely rapid at room temperature.   These reactions occur mainly in the gas or liquid state and include complexation, neutralization, and many oxidation-reduction reactions.   Fortunately, most reactions influencing the present condition of natural waters fall in this class.   There are, of course, notable exceptions which will be discussed later in conjunction with specific systems.

By combining our knowledge of reaction rates and thermodynamic stability we can often say a great deal about the conditions prevalent when a geologic system formed.   The coexistence of two phases which are thermodynamically unstable with respect to reaction to form a third phase can be used to place an upper limit on the temperature of the system when these phases were formed if we also know the dependence of the reaction rate on temperature.   The coexistence of unstable phases with their stable reaction product can sometimes be indicative of the duration of temperature and pressure conditions if we independently know these T, P conditions and the rates of the reactions involved.

We shall first develop the thermodynamic theory necessary for analyzing geological systems.   The latter part of this book will be devoted to applying this theory to specific natural chemical processes. Some of the questions to be considered in detail are:

1.   In a given mixture of gases such as CO_2, H_2O, and CH_4, held at a specified temperature and pressure, what would be the chemical composition of the equilibrium system?
2.   Under what conditions of temperature and pressure will calcite, the hexagonal form of $CaCO_3$, invert to aragonite, the orthorhombic form?
3.   What can the distribution pattern of trace elements and isotopes between coexisting mineral phases tell us about the conditions under which the minerals were formed?
4.   What factors influence the chemical composition of stream water?
5.   Can the distribution of a trace element, such as Ni, be used to determine the temperature of crystallization of a rock?
6.   How do chemical and physical properties of solids affect the structure of phase diagrams?

7. Why do some solids unmix, and how can we quantitatively discuss unmixing phenomena?

8. How do solutions at high temperatures and high pressures differ from those at room temperature and pressure?

We will be concerned with two distinct types of equilibria in our discussions. The first, and probably simplest, is mechanical equilibrium. This is of importance in dealing with polymorphic conversions such as calcite ↔ aragonite. In this case chemical reactions may be possible, but their rate is negligible with respect to the process being studied. The system would then be thermodynamically unstable as a whole but in a transient state of mechanical equilibrium. In most systems we will be concerned with chemical, or reaction, equilibrium. In these cases mechanical equilibrium may, or may not, be achieved. However, if chemical equilibrium is reached, mechanical equilibrium almost invariably is also achieved.

We shall begin by considering simple systems at equilibrium and attempt to define empirical and theoretical equations of state from the pressure and temperature dependence of the volume which these systems occupy.

There are many excellent textbooks of physical chemistry and thermodynamics. Books designed for chemists differ in their mode of presentation and range of topics from books designed for physicists and for engineers. The geologist will find useful information in each of these approaches. In general, books written for physicists and engineers contain a more thorough treatment of the solid state, whereas books for chemists deal predominantly with gases and liquids. The references listed at the end of this chapter contain samples of both types of presentation. The reader will undoubtedly find it beneficial to use these references to supplement the material presented in Chaps. 2 to 6.

Some of the problems given at the end of each chapter require experimental data not given in the text. Such problems are marked with an asterisk. Most information required may be found in the sources listed at the end of this chapter.

REFERENCES

Denbigh, Kenneth: "The Principles of Chemical Equilibrium," Cambridge University Press, New York, 1964.

Moore, Walter J.: "Physical Chemistry," Prentice-Hall, Inc., Englewood Cliffs, N.J., 1965.

Pitzer, Kenneth L., and Leo Brewer: "Thermodynamics," 2d ed., revision of Lewis and Randall, "Thermodynamics," 1st ed., McGraw-Hill Book Company, New York, 1961.

Swalin, Richard A.: "Thermodynamics of Solids," John Wiley & Sons, Inc., New York, 1962.

Zemansky, Mark W.: "Heat and Thermodynamics," 5th ed., McGraw-Hill Book Company, New York, 1968.

SOME SOURCES OF THERMODYNAMIC DATA

Clark, S. P., Jr. (ed.): Handbook of Physical Constants, *Geol. Soc. Am. Mem.* 97, 1966.

Garrels, R. M., and C. L. Christ: "Solutions, Minerals, and Equilibria," Harper & Row, Publishers, Incorporated, New York, 1965.

"Handbook of Chemistry and Physics," The Chemical Rubber Publishing Company, Cleveland, Ohio.

Kelley, K. K., and E. G. King: *U.S. Bur. Mines Bull.* 592, 1961.

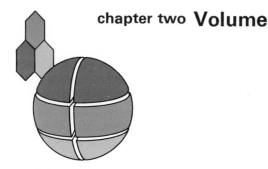

chapter two Volume

One of the factors influencing the nature of the equilibrium state of a chemical system is the volume which it occupies. Since different atomic configurations in general have significantly different densities, phase changes or chemical reactions cause volume changes. Further, the volume occupied by any given atomic configuration will change with temperature and with pressure. Our objective in this chapter will be to develop equations of state for various types of matter which will enable us to evaluate the volume changes resulting from changes in environmental conditions.

For a great variety of substances, the volume change of a system in response to changes in its environment can be completely specified in terms of temperature and pressure effects. The general form of the equation describing a volume change is

$$dV = \left(\frac{\partial V}{\partial T}\right)_P dT + \left(\frac{\partial V}{\partial P}\right)_T dP$$

This equation can also be written

$$dV = \alpha V \, dT - \beta V \, dP$$

where α is the coefficient of thermal expansion (the fractional change in the volume of a substance per degree increase in temperature at constant pressure), and β is the isothermal compressibility (the fractional change in volume per unit increase in pressure at constant temperature). In partial-derivative form,

$$\alpha = \frac{1}{V}\left(\frac{\partial V}{\partial T}\right)_P$$

$$\beta = -\frac{1}{V}\left(\frac{\partial V}{\partial P}\right)_T$$

Although we will generally use the compressibility, β, in this book, the reader should be aware that its reciprocal, B_T, the isothermal bulk modulus, is often used.   Hence

$$B_T = \frac{1}{\beta}$$

Our problem has now been reduced to finding for all substances of interest the values of α and β (or B_T) for all values of temperature and pressure.

EQUATION OF STATE FOR GASES

An extremely simple kinetic treatment leads to an equation of state which is valid for most gases at low pressures.   The assumptions made about the "ideal gas" are:

1. The gas consists of a large number of independent particles which occupy a negligible fraction of the total available space.
2. The particles obey Newton's laws of motion.
3. No electrostatic attractions exist between the particles.

The pressure exerted by such a gas on its container is proportional to the number of collisions which the molecules make against the container walls per unit time and to the average momentum change per collision.

Consider a cubic vessel containing Avogadro's number of molecules (1 mole of gas).   Although the actual molecular motion is chaotic, with each molecule frequently changing direction and velocity as the result of collisions, for the purposes of calculating the pressure exerted by the gas a macroscopically equivalent system can be adopted. In this system, one-third of the molecules travel in a direction perpendicular to each pair of container faces, and all the molecules travel with the mean velocity, $\bar{v}$. The pressure exerted by this organized gas will be the same as that of the random gas.

The number of impacts per unit area and time against a given container face is

$$\frac{N_0}{3} \frac{\bar{v}}{2l} \frac{1}{A}$$

where N_0 is Avogadro's number, 6.02×10^{23}; A is the area of the face; and l is the distance between container faces. The momentum change associated with each impact is $2m\bar{v}$, where m is the molecular mass. Since pressure is the rate of change of momentum per unit area, it is given by

$$P = \frac{N_0\bar{v}}{6lA} 2m\bar{v}$$

$$= \frac{N_0\bar{v}^2 m}{3V}$$

where V is the volume (lA) of the container and hence also of the gas. Furthermore, since $m\bar{v}^2/2$ is the mean kinetic energy per molecule, ϵ,

$$PV = \tfrac{2}{3}N_0\epsilon$$

The important result of this calculation is that, if the energy content of the gas, $N_0\epsilon$, is held constant, the volume of the gas will be inversely proportional to the pressure exerted on it.

The energy associated with translational movement in an ideal gas rises linearly with temperature, indicating that $N_0\epsilon$ is proportional to T. If $N_0\epsilon$ is proportional to T, the product, PV, must necessarily also be proportional to T.

If the constant of proportionality is designated R, then the equation of state for 1 mole of an ideal gas becomes

$$PV = RT$$

The value of this constant has been established experimentally. Its value in various units is as follows:

R = 1.986 cal/deg-mole
    = 0.08206 liter-atm/deg-mole
    = 8.314 joules/deg-mole = 8.314×10^7 ergs/deg-mole

It is useful to calculate the energy increment which must be added to each molecule in the gas in order to raise the temperature of the gas by one centigrade degree.   From the theory given above,

$$\tfrac{2}{3} N_0 \epsilon = RT$$

or

$$\epsilon = \frac{3}{2} \frac{R}{N_0} T$$

The ratio, R/N_0, of the universal gas constant to Avogadro's number is designated k (Boltzmann's constant).   The value of k generally used is that in the cgs system of units,

$$k = 1.380 \times 10^{-16} \text{ erg/deg}$$

If we differentiate the energy equation with respect to temperature, the result is

$$\frac{d\epsilon}{dT} = \frac{3}{2} k$$

Thus, for each degree of temperature increase, an amount of translational energy equal to $\tfrac{3}{2}k$ must be added to each molecule of gas.

The ideal gas equation of state can be put into the general form given above if α and β are computed:

$$\alpha = \frac{1}{V} \left(\frac{\partial}{\partial T} \frac{RT}{P} \right)_P = \frac{R}{VP} = \frac{1}{T}$$

$$\beta = -\frac{1}{V} \left(\frac{\partial}{\partial P} \frac{RT}{P} \right)_T = \frac{RT}{VP^2} = \frac{1}{P}$$

It should be noted that the bulk modulus, B_T, of an ideal gas is simply its pressure.   Substituting the calculated values of α and β into the volume equation gives

$$dV = \frac{V}{T} dT - \frac{V}{P} dP$$

This equation can be rewritten

$$\frac{dV}{V} = \frac{dT}{T} - \frac{dP}{P}$$

In this form the ideal gas law states that the fractional increase in volume experienced by a gas as the result of a change in environmental conditions is equal to the fractional increase in absolute temperature minus the fractional increase in pressure.

Real gas behavior does not always follow the ideal gas law described above. At low pressures, where the mean distance separating the molecules in a gas is hundreds or more times the molecular diameter, the space occupied by the molecules is truly negligible. However, as the pressure increases, the volume occupied by the molecules becomes a sizable fraction of the total volume. Similarly, the mutual electrostatic force between molecules has little effect when the molecules are well separated but becomes important as the gas becomes more dense. Thus, although the ideal gas law holds for all gases in the limit of vanishingly small pressure, and for most gases up to moderate pressures of several atmospheres, we must seek elsewhere for an equation of state for real gases at elevated pressures.

Van der Waals proposed an improved equation of state which takes into account the finite molecular volume and the mutual attraction between molecules. Whereas the ideal gas law states

$$P = \frac{RT}{V}$$

the van der Waals equation states

$$P = \frac{RT}{V - b} - \frac{a}{V^2}$$

Elimination of the molecular-volume effect is easily accomplished if the total molar volume, V, is replaced by the free volume, $V - b$. The constant, b, is the volume made unavailable for movement because of the finite size of the molecules. If this were the only difference between a real gas and an ideal gas, the pressure predicted by the ideal gas law would show increasing deviation from the actual pressure as the density of the gas was increased. At elevated pressures a real gas would exert a higher pressure than that predicted by the ideal gas law. In terms of the kinetic model, the exclusion of volume increases the frequency with which gas molecules strike the walls of the container. This is most easily seen by considering two molecules traveling perpendicular to the same wall of a rectangular chamber. If the molecules were not on a collision course, the number of impacts with each wall would be $\bar{v}$, the average velocity of the molecules, divided by l, the distance between the walls. If the molecules were to collide, each would travel a distance shorter by twice the molecular radius, r; hence the collision frequency would rise to $\bar{v}/(l - 2r)$.

We can obtain an estimate of the constant, b, by considering a hypothetical collision between two molecules. As shown in Fig. 2.1, each molecule excludes a volume equal to that of a sphere with twice its radius. The excluded volume is 8 times the volume of the molecule itself. However, since each collision involves a *pair* of molecules, the volume forbidden to each molecule is only 4 times the molecular volume. The constant, b, should be 4 times the volume of Avogadro's number of molecules. Since most gas molecules are of the order of 1.5×10^{-8} cm in radius, b should average 0.035 liter/mole. The prediction agrees reasonably well with experimentally determined values of b (see Table 2.1).

The molecular-attraction effect causes a reduction in the pressure which a gas exerts on its container. Within a gas these attractions are equal in all directions and hence have no net effect on the molecular motion. Near the walls of the container, the forces pull in only one direction. A molecule attempting to leave the gas and hit the wall will be "restrained" by its fellow molecules. The velocity of all molecules near the edge of the container will be decreased, causing a reduced force of impact and commensurately lower pressure. Some molecules will lack the kinetic energy to "escape" the gas and will fall back into the gas, much as a rocket lacking escape velocity falls back to the earth.

The mutual attraction between molecules in a gas results to a first approximation from electrostatic interaction between dipoles. All molecules have nonuniform charge distribution; that is, they are more negative on one side than the other. This nonuniformity may be perma-

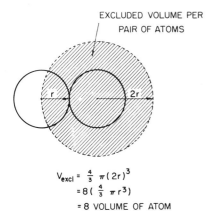

EXCLUDED VOLUME PER
PAIR OF ATOMS

$$V_{excl} = \frac{4}{3} \pi (2r)^3$$
$$= 8\left(\frac{4}{3} \pi r^3\right)$$
$$= 8 \text{ VOLUME OF ATOM}$$

Fig. 2.1. Mechanical-collision model to show the effect of the finite volume of molecules on the pressure of a real gas.

TABLE 2.1

Van der Waals Constants (a and b),† Critical Conditions (T_c, V_c, P_c), and Boyle Temperature (T_B) for a Number of Gases‡

Gas	a, liter^2-atm/mole^2	b, liters/mole	T_c, °K	V_c, cm^3/mole	P_c, atm	T_B $= a/Rb$
Helium	0.03412	0.02370	5.2	61.55	2.25	17.5
Neon	0.2107	0.01709	44.75	44.30	26.86	150
Argon	1.345	0.03219	150.87	74.56	48.34	508
Krypton	2.318	0.03978	209.39	92.08	45.182	710
Xenon	4.194	0.05105	289.9	118.8	58.2	1000
Hydrogen	0.2444	0.02661	33.2	69.68	12.8	112
Oxygen	1.360	0.03183	154.28	74.42	49.713	521
Nitrogen	1.390	0.03913	125.97	90.03	33.49	433
Chlorine	6.493	0.05622	417.1	123.4	76.1	1410
Carbon monoxide	1.485	0.03985	134.4	90.03	34.6	455
Carbon dioxide	3.592	0.04267	304.16	94.23	72.83	1025
Water	5.464	0.03049	647.3	55.44	218.5	2180
Sulfur dioxide	6.714	0.05636	430.25	124.8	77.65	1450
Carbon disulfide	11.62	0.07685	546.15	172.7	72.868	1845
Ammonia	4.170	0.03707	405.5	72.02	112.2	1370
Carbon tetra-chloride	20.39	0.1383	556.25	275.8	44.98	17950
Benzene	18.00	0.1154	561.6	256.4	47.89	19000

† Van der Waals constants taken from "Handbook of Chemistry and Physics," 45th ed., The Chemical Rubber Publishing Company, Cleveland, Ohio.
‡ Critical-point data taken from E. A. Moelwyn-Hughes, "Physical Chemistry," 2d ed., p. 586, Pergamon Press, New York, 1961.

nent as in the case of H_2O, resulting from the asymmetry of the molecule, or statistical as in the case of Ar, where it arises from random fluctuations in the positions of the electrons about the nucleus. Although the repulsions between oppositely oriented molecules tend to compensate for the attractions between like-oriented molecules, the tendency toward dipole alignment leads to a net attractive force. In the case of oppositely charged molecules, or ion pairs, the attractive force varies inversely with the square of the distance of separation, and pressure varies inversely with the fourth power of this distance. For dipole interactions, the force drops with the fourth power of the distance; hence pressure depends on the sixth power. The decrease in pressure in the gas due to electrostatic effect falls with the square of the volume (as $V^2 \propto r^6$). The constant, a, is a measure of the magnitude of the dipole moment of the molecules. As shown in Table 2.1, it is fairly large for asymmetric molecules like H_2O and CS_2 and rather small for the rare gases He and Ne.

It is convenient to think of the pressure terms in the van der Waals equation in the following manner. The molecules making up the gas are held apart by a thermal pressure, $RT/(V - b)$. This pressure is balanced by (1) the pressure resulting from the electrostatic attraction between the molecules, a/V^2, the internal pressure, and (2) the external pressure, P. Hence there is an equilibrium between various kinds of forces:

Pressure holding molecules apart = pressure pushing molecules together

$$\frac{RT}{V - b} = \frac{a}{V^2} + P$$

The terms *internal, external,* and *thermal pressure* will be used frequently in the discussions which follow.

Since the two causes of deviation from the ideal pressure exerted by a gas act in opposite directions, we might expect some set of conditions under which the opposing effects cancel each other. To determine these conditions, we define the compressibility factor

$$Z = \frac{PV}{RT}$$

For an ideal gas, Z is unity. For a van der Waals gas,

$$Z = \frac{[RT/(V - b) - a/V^2]V}{RT}$$

$$= \frac{1}{1 - b/V} - \frac{a}{RTV}$$

Since b/V is generally much smaller than unity, we may expand the first term of this equation in series, giving

$$Z \cong 1 + \left(b - \frac{a}{RT}\right)\frac{1}{V} + \left(\frac{b}{V}\right)^2 + \left(\frac{b}{V}\right)^3 + \cdots$$

Z is equal to unity over an extended range of pressure if the second term of the expansion is zero. This is true when

$$T_B = \frac{a}{Rb}$$

This temperature, at which the gas behaves ideally over a wide range of pressure, is called the *Boyle temperature* of the gas. A plot of the compressibility factor versus pressure is given for CH_4 in Fig. 2.2. As can be seen from the graph, considerable deviation from ideality occurs at moderate pressures when the temperature is much higher or lower than the Boyle temperature. At or near the Boyle temperature the gas behaves ideally up to fairly high pressures. Since b is nearly the same for most gases, T_B is determined mainly by a. Gases consisting of asymmetric molecules have higher Boyle temperatures than those with symmetric molecules (see Table 2.1).

For temperatures greater than the Boyle temperature the volume effect dominates. The observed pressures are always greater than those predicted by the ideal gas law for the same temperature and volume.

The reason for this can be easily shown. If the volume effect alone were operative,

$$Z = \frac{[RT/(V-b)]V}{RT} = \frac{V}{V-b}$$

The amount by which Z exceeds unity depends only on the molecular volume occupied by the gas, not on its temperature. On the other hand, if the electrostatic effect alone were operative,

$$Z = \frac{(RT/V - a/V^2)V}{RT} = 1 - \frac{a}{RVT}$$

The amount by which Z falls below unity depends not only on V but also on T. As T rises, at any given value of the volume or gas density

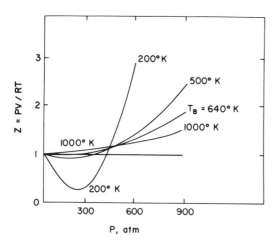

Fig. 2.2. Compressibility factor, Z, plotted as a function of pressure for the gas CH_4. *(After Castellan.)*

the molecular-volume effect will not change but the electrostatic effect
will decrease. T_B is the temperature at which the two effects become
equal. Above T_B the volume effect dominates at all pressures.

For temperatures below the Boyle temperature the molecular-
attraction effect dominates at low densities, but as the density increases,
the excluded-volume effect becomes more and more important. Eventually, at high densities it becomes the dominant cause of nonideality.
For a given temperature, the molar volume, V_B, for which the compressibility factor crosses unity is given by

$$V_B \cong b\,\frac{T_B}{T} - 1$$

Although differences in molecular size and degree of charge asymmetry lead to a wide range of molar volumes for different gases contained at the same temperature and pressure, it is possible to demonstrate a uniformity in the character of the nonideality of these gases.
This is done by normalizing both temperature and pressure to some
multiple of those values observed when the gas shows a given degree of
nonideality. The degree of nonideality chosen for this comparison is
that found at the so-called critical point.

The nature of this point can be seen as follows: When a liquid is
heated in a closed container, its vapor pressure increases with increasing
temperature. Concurrently, the density of the liquid phase decreases
and that of the vapor phase increases. Similarly, other properties of
the gas and liquid approach each other as the temperature increases.
Eventually, all the properties of the liquid and gas phases become identical. The temperature at which this happens is called the *critical
temperature*, T_c; the pressure at the critical point is P_c, the *critical
pressure*. The ratios of the variables P and T to the critical constants,
P_c and T_c, for that gas are called the *reduced variables* of the gas:

$$P_r = \frac{P}{P_c} \qquad \text{and} \qquad T_r = \frac{T}{T_c}$$

Similarly, V_c is the critical volume, and $V_r = V/V_c$ is the reduced volume. Two gases which have the same values of their *reduced* variables
are said to be in *corresponding states*. To a close approximation, most
gases at moderate pressures show the same density when at the same
T_r and P_r.

As an example of an equation of state using reduced parameters,
the van der Waals equation can be converted to this form. In order to
make this conversion, we must examine the form of the van der Waals

equation more closely.   Figure 2.3 shows the P-V isotherms for CO_2 near the critical point.   The bell-shaped dashed line outlines the "two-phase region" where liquid and gaseous CO_2 coexist in equilibrium. To the right of the two-phase field, only the gas is present, and to the left, only the liquid is present.

The dashed lines inside the two-phase field are the solutions to the van der Waals equation at two different fixed temperatures.   Since the van der Waals equation cannot describe two coexisting phases, the calculated maximum and minimum shown are not physically meaningful.   However, as higher-temperature isotherms are calculated, the maximum and minimum approach each other and merge into an inflec-

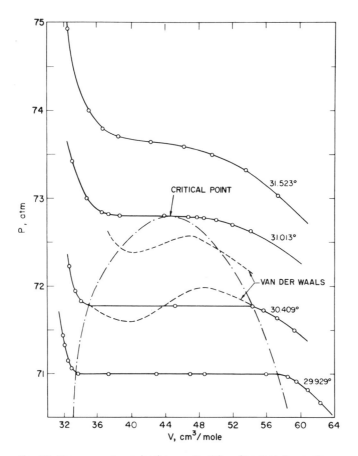

Fig. 2.3. Pressure-volume isotherms for CO_2. See text for explanation of dashed lines. *(After Walter J. Moore, "Physical Chemistry," Prentice-Hall, Inc., Englewood Cliffs, N.J., 1965.)*

tion point at the critical temperature. At the inflection point both the first and second derivatives of pressure with respect to volume at constant temperature must be zero.

$$\left(\frac{\partial P}{\partial V}\right)_T = 0 \quad \text{and} \quad \left(\frac{\partial^2 P}{\partial V^2}\right)_T = 0$$

Therefore, at the critical point the following equations must hold:

$$P_c = \frac{RT_c}{V_c - b} - \frac{a}{V_c^2}$$

$$\left(\frac{\partial P}{\partial V}\right)_T = -\frac{RT_c}{(V_c - b)^2} + \frac{2a}{V_c^3} = 0$$

$$\left(\frac{\partial^2 P}{\partial V^2}\right)_T = \frac{2RT_c}{(V_c - b)^3} - \frac{6a}{V_c^4} = 0$$

These equations can be solved to give the critical constants in terms of the van der Waals constants, a and b, and the gas constant, R:

$$P_c = \frac{a}{27b^2} \qquad V_c = 3b \qquad T_c = \frac{8a}{27bR}$$

In addition, it is easily shown that

$$\frac{RT_c}{P_c V_c} = \frac{8}{3} \quad \text{and} \quad T_B = \tfrac{27}{8} T_c$$

The variables of state, P, V, and T, may then be converted into reduced variables:

$$P = \frac{a}{27b^2} P_r \qquad V = 3b V_r \qquad T = \frac{8a}{27bR} T_r$$

Substituting into the van der Waals equation, we find

$$P_r = \frac{\frac{8}{3} T_r}{V_r - \frac{1}{3}} - \frac{3}{V_r^2}$$

The equation in this form contains only constants and reduced variables.

We can test the theory of corresponding states by plotting the compressibility factor, Z, against the reduced pressure, P_r, for several gases.

If the gases are truly in corresponding states, we should have one curve for each reduced temperature on which all the individual gas curves would lie. Figure 2.4 shows this plot for three gases with very different physical properties. The agreement with the law of corresponding states is remarkably good.

It is of interest to calculate the coefficient of thermal expansion, α, and the compressibility, β, for a van der Waals gas. When this is done, the following results are obtained:

$$\alpha = \frac{f}{T} \frac{1}{1 - [2f/(1 + P/P_{\text{int}})]}$$

and

$$\beta = \frac{f}{P + P_{\text{int}}} \frac{1}{1 - [2f/(1 + P/P_{\text{int}})]}$$

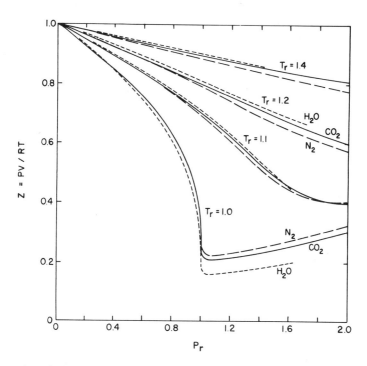

Fig. 2.4. Compressibility factor, Z, plotted against the reduced pressure, P_r, for several values of the reduced temperature. This plot shows the applicability of the law of corresponding states for gases. (*After Kenneth L. Pitzer and Leo Brewer, "Thermodynamics," 2d ed., revision of Lewis and Randall, "Thermodynamics," 1st ed., McGraw-Hill Book Company, New York, 1961.*)

where

$$f = \frac{V - b}{V} = \text{fraction of free volume}$$

and

$$P_{\text{int}} = \frac{a}{V^2} = \text{internal pressure}$$

Under most circumstances $2f/(1 + P/P_{\text{int}})$ for gases is much less than unity, which allows the approximations

$$\alpha \cong \frac{f}{T}$$

and

$$\beta \cong \frac{f}{P + P_{\text{int}}} = \frac{f}{P_{\text{thermal}}}$$

Since for an ideal gas $f = 1$ and $P_{\text{int}} = 0$, these expressions reduce to the ideal gas values. The van der Waals equivalent of the differential form of the ideal gas law becomes

$$\frac{dV}{V_{\text{free}}} = \frac{dT}{T} - \frac{dP}{P_{\text{thermal}}}$$

where $V_{\text{free}} = V - b$. For each percent the absolute temperature increases, the *free* volume increases by 1 percent, and for each percent the *thermal* pressure of the gas increases, the *free* volume decreases by 1 percent.

Written in the ideal gas form, the van der Waals equation becomes

$$P_{\text{thermal}} V_{\text{free}} = RT$$

Although the van der Waals equation proves extremely valuable in understanding density variations in real gases, as a practical equation of state it suffers from two gross difficulties. First, it fails to yield sufficiently accurate density estimates for elevated pressures, and, second, prohibitive algebraic difficulties are encountered in combining the van der Waals equation with other thermodynamic equations.

Because of these difficulties it is necessary, as well as convenient, to use the so-called virial equation of state:

$$PV = A + BP + CP^2 + DP^3 + \cdots$$

A, B, C, and D are the virial coefficients and must be evaluated for each gas at each temperature of interest. However, once the virial coefficients have been determined, this equation is easier to use than any of the theoretical equations which allow for nonideal behavior. It also offers any degree of accuracy required (provided, of course, sufficiently accurate experimental P, V, T data are available for the determination of the virial coefficients). At low pressures, it is often possible to use a virial equation of the simplified form

$$PV = RT + BP$$

Tables 2.2 and 2.3 show the gas molar volume as a function of temperature and pressure for H_2O and CO_2. The second column was

TABLE 2.2

Molar Volume of Water Vapor

Conditions[†]	Ideal Gas, liters	Van der Waals, liters	Experi-mental,[‡] liters	V[¶] $- b$, liters	a/V^2, atm
1 bar, 120°C	32.69	32.55	32.32	32.52	0.00516
1 bar, 200°C	39.34	39.23	39.11	39.20	0.00355
1 bar, 400°C	55.97	55.90	55.89	55.87	0.00175
1 bar, 600°C	72.60	72.56	72.56	72.53	0.00104
1 bar, 800°C	89.23	89.20	89.19	89.17	0.00069
1 bar, 1000°C	105.86	105.84	105.84	105.81	0.00049
10 bars, 200°C	3.934	3.821	3.726	3.791	0.374
10 bars, 400°C	5.597	5.527	5.523	5.497	0.179
10 bars, 600°C	7.260	7.214	7.228	7.184	0.105
10 bars, 800°C	8.923	8.891	8.899	8.861	0.069
10 bars, 1000°C	10.586	10.563	10.580	10.533	0.049
100 bars, 400°C	0.5597	0.4828	0.4760	0.4523	23.44
100 bars, 600°C	0.7260	0.6786	0.6896	0.6481	11.86
100 bars, 800°C	0.8923	0.8608	0.8744	0.8303	7.37
100 bars, 1000°C	1.0586	1.0372	1.0519	1.0068	5.08
200 bars, 400°C	0.2799	0.1858	0.1795	0.1553	158.28
200 bars, 600°C	0.3630	0.3139	0.3270	0.2834	55.46
200 bars, 800°C	0.4462	0.4148	0.4293	0.3843	31.77
200 bars, 1000°C	0.5293	0.5087	0.5239	0.4782	21.11

[†] 1 bar = 0.9869 atm.
[‡] Data taken from S. P. Clark, Jr. (ed.), Handbook of Physical Constants, sec. 16, *Geol. Soc. Am. Mem.* 97, 1966.
[¶] Van der Waals volume.

calculated by using the ideal gas law; the third by the van der Waals equation. Column 4 is the experimental determination. It is evident that the van der Waals equation is valid over a much larger range than the ideal gas law. Only at low pressure and high temperature does the ideal gas law give very good results.

TABLE 2.3
Molar Volume of Carbon Dioxide

Conditions†	Ideal Gas, liters	Van der Waals, liters	Experimental,‡ liters	$V\P - b$, liters	a/V^2, atm
25 bars, 200°C	1.574	1.524	1.528	1.481	1.55
25 bars, 400°C	2.239	2.218	2.234	2.175	0.730
25 bars, 800°C	3.570	3.573	3.607	3.530	0.281
100 bars, 200°C	0.393	0.343	0.361	0.301	30.49
100 bars, 400°C	0.560	0.540	0.558	0.500	12.30
100 bars, 800°C	0.892	0.896	0.909	0.854	4.47
200 bars, 400°C	0.280	0.265	0.281	0.222	51.19
200 bars, 800°C	0.446	0.452	0.464	0.410	17.55
400 bars, 800°C	0.223	0.234	0.244	0.191	65.60

† 1 bar = 0.9869 atm.
‡ Data taken from S. P. Clark, Jr. (ed.), Handbook of Physical Constants, sec. 16, *Geol. Soc. Am. Mem.* 97, 1966.
¶ Van der Waals volume.

EQUATIONS OF STATE FOR LIQUIDS

Liquids are nearly incompressible when compared with gases. At 25°C the volume of a gas would halve if the external pressure were raised from 1 to 2 atm. The same change would produce only a 50 parts per million (ppm) decrease in the volume of liquid water. At 1 atm a gas would expand by 1 percent when heated from 25 to 28°C. Water would expand only 0.06 percent. Despite their small magnitude, the compressibilities and coefficients of thermal expansion are important to problems in chemical equilibrium. We will therefore explore the factors controlling their magnitudes and their pressure and temperature dependences.

One might think that a liquid should resemble a high-density gas. If this were the case, then α should be given by some function of the fraction of the volume not occupied by the molecules, X, divided by the absolute temperature. Hence

$$\alpha = \frac{X}{T}$$

In like manner, β should be the same function of the fraction of free volume divided by the thermal pressure holding the molecules apart. This thermal pressure is the sum of the internal and external pressures acting on the liquid. Hence

$$\beta = \frac{X}{P_{\text{thermal}}} = \frac{X}{P_{\text{ext}} + P_{\text{int}}}$$

As stated above for a van der Waals gas,

$$\alpha = \frac{f}{T} \frac{1}{1 - [2f/(1 + P/P_{\text{int}})]} = \frac{X}{T}$$

and

$$\beta = \frac{f}{P + P_{\text{int}}} \frac{1}{1 - [2f/(1 + P/P_{\text{int}})]} = \frac{X}{P_{\text{thermal}}}$$

where

$$X = f \frac{1}{1 - [2f/(1 + P/P_{\text{int}})]}$$

It is evident that, when the internal pressure is much greater than the external pressure, X is simply $f/(1 - 2f)$.

Tait has shown that the compressibility of water follows the empirical equation

$$\beta = \frac{A}{P + B} = \frac{0.137}{P + 2800} \qquad \text{atm}^{-1}$$

where A and B are constants independent of pressure. This relationship is presented graphically in Fig. 2.5. The form of this equation is in accordance with that predicted by the van der Waals equation. The Tait constant, B, is equivalent to the internal pressure, a/V^2, and his constant A, to the parameter X.

For pressures up to 1000 atm neither the empirically determined A nor B change appreciably. That this is again in agreement with the van der Waals prediction is shown as follows: Over this pressure range

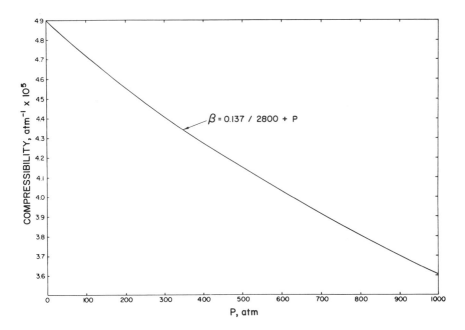

Fig. 2.5. Tait equation for H_2O at room temperature.

the volume, V, of H_2O changes about 4 percent. Hence

$$\frac{B_{1000}}{B_1} = \left(\frac{V_1}{V_{1000}}\right)^2 = (1.04)^2 = 1.08$$

The value of f at 1 atm can be calculated from the Tait parameter A as follows:

$$A = X \cong \frac{f}{1 - 2f}$$

Hence

$$f = \frac{A}{1 + 2A} = \frac{0.137}{1.274} \cong 0.107$$

It follows that

$$\frac{A_{1000}}{A_1} = \frac{1 - [0.214/(1 + 1/2800)]}{1 - [0.214/(1 + 1000/2800)]} = \frac{1 - 0.214}{1 - 0.158} = 0.935$$

One might ask whether the magnitude of the internal pressure for H_2O calculated from the van der Waals equation agrees with Tait's constant of 2800 atm. Since at low pressures a/V^2 must essentially equal

$RT/(V - b)$, we have

$$B \cong \frac{RT}{fV}$$

$$= \frac{(0.082 \text{ liter-atm/deg-mole})(300 \text{ deg})}{(0.107)(0.018 \text{ liter/mole})}$$

$$= 12{,}800 \text{ atm}$$

This result is far greater than the empirical value. However, H_2O molecules are known to associate in the liquid phase so that the mean molecular weight is about 60 instead of 18. If this is taken into account, the predicted internal pressure falls to 3800 atm, a value much more nearly in agreement with Tait's 2800-atm value.

The van der Waals theory predicts a value of

$$\alpha = \frac{X}{T} = \frac{0.137}{300} = 4.6 \times 10^{-4} \text{ deg}^{-1}$$

for the coefficient of thermal expansion of H_2O. The observed room-temperature value of α is 2×10^{-4} deg^{-1}. It is clear that the van der Waals theory is not capable of providing accurate predictions of α and β; however, it does provide a basis for understanding their magnitudes and their temperature and pressure dependences.

EQUATIONS OF STATE FOR SOLIDS

For solids there is no justification for the use of the van der Waals equation of state. The atoms are in fixed positions and hence do not undergo the random translational motions which characterize gases. A new model is therefore necessary.

The simplest solids to treat are those bound by electrostatic attraction between ions. The alkali halide salts approach this ideal. In KCl (the mineral sylvite), for example, each K atom is stripped of one electron and each Cl atom has one extra electron. The positively charged K atoms attract the negatively charged Cl atoms with a force, F, proportional to the product of their charges divided by the square of the distance separating their centers. Of course, there is a corresponding negative force of repulsion between Cl ions and between K ions. Because in sylvite the six nearest neighbors of any given Cl$^-$ ion are K$^+$ ions, and vice versa, the sum of the positive forces exceeds that of the negative forces.

In addition to the net force of electrostatic attraction between the ions there is also a powerful short-range force of repulsion which operates between "touching" ions.   Since the latter force changes very rapidly with distance, the ions have an almost rigid character.   The electrostatic forces pull the adjacent ions together until this "wall of repulsion" prevents further collapse.   The wall of repulsion arises from the interaction between individual electrons when two ions are brought sufficiently close to create serious overlap between their respective electron clouds. The force of repulsion can be approximated by

$$F = \frac{B}{x^n}$$

where n is about 10 for alkali halide compounds.

If a single K^+ and a single Cl^- ion were brought together, they would come to rest when the repulsive force exactly balanced the attractive force, or when

$$\frac{e^2}{x^2} = \frac{B}{x^n}$$

If the pair is subjected to mechanical squeezing pressure, P, taking the cross-sectional area over which the pressure acts to be the ionic diameter squared, hence approximately x^2, we have

$$Px^2 + \frac{e^2}{x^2} = \frac{B}{x^n}$$

Thus

$$P = \frac{B}{x^{n+2}} - \frac{e^2}{x^4}$$

The fractional change in separation per unit of added external pressure, $-(1/x)(dx/dP)$, is given by

$$-\frac{1}{x(dP/dx)} = \frac{1}{(n+2)[B/(x^{n+2})] - 4e^2/x^4}$$

However, since

$$\frac{B}{x^{n+2}} = P + \frac{e^2}{x^4}$$

then

$$-\frac{1}{x}\frac{dx}{dP} = \frac{1/(n+2)}{P + [(n-2)/(n+2)](e^2/x^4)}$$

Taking $x = 3 \times 10^{-8}$ cm, $e = 4.8 \times 10^{-10}$ esu, $n = 10$, and 10^6 esu^2/ cm^4 = 1 atm, we have

$$-\frac{1}{x}\frac{dx}{dP} = \frac{0.08}{P + (0.16 \times 10^6)}$$

Thus at low pressures the separation distance should decrease $\frac{1}{2}$ ppm for each atmosphere of pressure exerted.

Although for a real crystal the compressibility is more difficult to calculate, we can arrive at an approximate result quickly by using two additional facts. First, if all the bonds in the crystal are shortened by an amount dx/x then the volume must diminish by an amount $3dx/x$. Second, the vector sum of all the attractive and repulsive forces acting between a single K$^+$-Cl$^-$ pair is 1.7 times that for the lone ion pair. Hence

$$\beta = -\frac{1}{V}\frac{dV}{dP} = -\frac{3}{x}\frac{dx}{dP} = \frac{3/(n+2)}{P + 1.7[(n-2)/(n+2)](e^2/x^4)}$$
$$= \frac{0.25}{P + (0.27 \times 10^6)}$$

or about 1-ppm decrease in volume per atmosphere increase in pressure. The experimental value is 5 ppm.

The inverse of β gives the theoretical isothermal bulk modulus, B_T, of KCl. At low pressure we would have

$$B_T = \frac{0.27 \times 10^6}{0.25} \text{ atm} \approx 1000 \text{ kbars}$$

This is approximately a factor of 5 higher than the experimental value of B_T.

It is interesting to note that again the compressibility comes out in the Tait form. The internal pressure is given by

$$\frac{1.7(n-2)e^2}{(n+2)x^4}$$

which is 2.7×10^5 atm for KCl. The fraction of free volume is

$$\frac{3/(n+2)}{1 + 6/(n+2)}$$

which for KCl is 0.167.

We have not yet considered the vibrations which occur in solids. The addition of vibrational energy gives rise to the expansion of solids in a manner similar to the expansion of gases by addition of translational energy.   The reason for volume expansivity of solids is not obvious.   If, for example, the vibrations were symmetric about the center of mass, the volume of the solid would remain unchanged with increasing temperature. The presence of thermal expansivity is a result of the asymmetric nature of the forces between atoms.

Figure 2.6 is a schematic potential diagram, showing potential energy, φ, as a function of the distance between atoms.   In our simple case the potential energy is that function which can be differentiated with respect to distance to give the force.   Thus,

$$-\frac{d\varphi}{dx} = F$$

If the potential energy is less than or equal to zero for a diatomic molecule, there is a chemical bond between the two atoms.   If the potential is greater than zero, the two atoms exist separately.

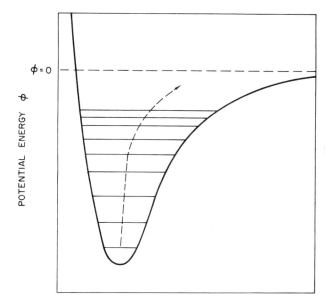

DISTANCE OF SEPARATION

Fig. 2.6. Schematic potential diagram for a diatomic oscillator. Horizontal lines in the "potential well" represent vibrational states. The dashed line shows the equilibrium distance of separation for the two atoms.

The potential is the sum of two terms, one for attractive potential proportional to $1/x$, and one for repulsive potential proportional to $1/(x^n - 1)$. This leads to an asymmetric total potential. The dashed line in Fig. 2.6 connects the equilibrium position of the atoms in each vibrational state. As potential energy is increased, the equilibrium distance of separation increases and the solid expands.

As we shall see in the next chapter, the heat capacity of a solid is a measure of energy taken up in vibrations. The temperature dependence of the coefficient of thermal expansion should thus be related to the temperature dependence of heat capacity. As we shall see, both heat capacity and the coefficient of thermal expansion are zero at absolute zero and rise toward a nearly constant value at elevated temperatures.

The ionic theory predicts a rather definite relationship between compressibility and mean volume per ion pair for compounds of similar structure and cation-anion valence. If the value of the repulsive exponent, n, for a group of these compounds is nearly the same, the reciprocal of the compressibility should vary linearly with $V^{\frac{1}{3}}$. This can be seen as follows: As shown above,

$$\beta = \frac{3/(n + 2)}{P + A[(n - 2)/(n + 2)](e^2/x^4)}$$

where A is the product of the anion and cation charges, $Z_C \times Z_A$, multiplied by a geometric constant dependent on the atomic arrangement (that is, 1.7 for KCl). For values of $A[(n - 2)/(n + 2)](e^2/x^4)$ much greater than P we may approximate the bulk modulus by

$$B_T = \frac{1}{\beta} = \frac{A}{3}(n - 2)e^2 x^{-4}$$

Taking the logarithm of both sides and setting $\log[A(n - 2)e^2/3]$ equal to a constant, C, we have

$$\log B_T = -\tfrac{4}{3}\log V + C$$

This equation describes a family of straight lines, since A and n may be different for different compounds. For each pair of A and n values a unique straight line is described. Since chemically related compounds should have essentially the same values of A and n, chemical "families" which share the same valence and crystal structure should plot on the same line.

The log of the bulk modulus for numerous solid compounds is plotted against the log of the volume per ion pair in Fig. 2.7. As predicted, the

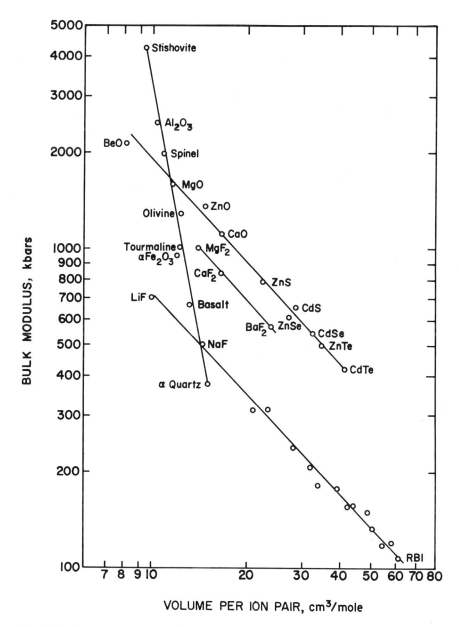

Fig. 2.7. Bulk modulus versus volume per ion pair for solid compounds. *(After Anderson and Soga.)*

compounds of similar valence and crystal structure fall along straight lines with slopes close to the predicted value. The actual slope of -1 can be derived by a more complete theoretical treatment.

Certain oxide compounds, notably those containing Si, Al, Mg, and possibly Fe, fall on a line with a slope of -4. Since almost all rocks fall on this same line, it is of great geologic significance. What is the origin of a line with a slope of -4? There is not at present complete agreement on the meaning of this linear arrangement of oxides; however, there are several features which can help us interpret the meaning of the line.

The first point to notice is that the high-pressure form of SiO_2, stishovite, and the low-pressure form, α quartz, lie on the line. This suggests that the effect of pressure on the compressibility might play an important part in our understanding of this relationship. For any substance, both its compressibility and volume per ion pair decrease with pressure. From the equations given above,

$$\frac{d(1/\beta)}{dP} = \frac{dB_T}{dP} = \frac{n+2}{3} + \frac{4}{3}$$

and

$$\frac{dV}{dP} = -\beta V = \frac{-V}{B_T}$$

Hence

$$\frac{dB_T}{dV} = \frac{dB_T}{dP}\frac{dP}{dV} = -\frac{n+6}{3}\frac{B_T}{V}$$

Integration gives

$$\ln B_T = -\frac{n+6}{3}\ln V + C$$

Since n is about 6 for oxides and silicates, the slope of this straight line should be about -4. Indeed, almost all ionic compounds, including the oxides, show this sort of pressure relationship between bulk modulus and volume. Our clue is that the reduction of the volume occupied per ion pair in going from oxide to oxide in certain cases yields large changes in compressibility similar to those obtained when atoms are forced closer together under great pressure, rather than the much smaller changes predicted if smaller ions were substituted for larger ones. This difference would result if the normal compounds were held apart largely by anion-cation contacts and these "anomalous" oxides were held apart largely by anion-anion contacts. If this were the case, substitution of a smaller ion in these oxides would not necessarily lead to a volume change. Those volume changes which occur reflect increased anion-cation forces which serve to increase the pressure on the oxygen-oxygen contacts. On the other hand, in nonoxygen compounds and "normal" oxides substitution of a smaller ion would permit a decrease in volume purely because more efficient packing is achieved.   In other words, in one case a lattice of touching oxygen atoms is forced together by the electrostatic pull of loosely fitting cations, and in the other, the anions are held apart by over-sized cations.   Pauling assumed that anions were never in contact in ionic compounds because the coordination number of cations always adjusted itself to prevent this from taking place.   Whereas Pauling's rule seems to work for many groups of compounds, including those which are highly covalent in character, it does not appear to hold for all the oxides.

The densities of a number of rock-forming oxides are given in Table 2.4.   The molar volume divided by the number of oxygen atoms in the formula is tabulated as the volume per oxygen atom.   The volume/oxygen varies from 7.0 for stishovite, the ultradense form of SiO_2, to 13.3 for tridymite, a low-pressure, high-temperature form of SiO_2.   All other oxide minerals fall between these extremes.   Table 2.5 shows the variation of the volume per oxygen atom for different mineral assemblages having the same chemical composition.

TABLE 2.4

Oxygen Packing in Some Common Oxide Minerals†

Mineral	Formula	Density	Molecular Weight	Molar‡ Volume	Volume‡ Atom	Volume‡ Oxygen
Tridymite	SiO_2	2.265	60.09	26.53	8.84	13.3
Cristobalite	SiO_2	2.334	60.09	25.74	8.55	12.8
Calcite	$CaCO_3$	2.712	100.09	36.94	7.39	12.3
Aragonite	$CaCO_3$	2.930	100.09	34.16	6.83	11.4
Quartz	SiO_2	2.648	60.09	22.69	7.56	11.3
Periclase	MgO	3.584	40.32	11.25	5.63	11.3
Olivine	Mg_2SiO_4	3.214	140.73	43.79	6.30	11.0
Enstatite	$MgSiO_3$	3.198	100.41	31.40	6.28	10.4
Andalusite	Al_2SiO_5	3.144	162.05	51.54	6.44	10.3
Coesite	SiO_2	2.911	60.09	20.64	6.88	10.3
Anatase	TiO_2	3.899	79.90	20.49	6.83	10.3
Sillimanite	Al_2SiO_5	3.247	162.05	49.91	6.24	10.0
Spinel	$MgAl_2O_4$	3.582	142.28	39.72	5.67	9.9
Brookite	TiO_2	4.123	79.90	19.38	6.46	9.7
Rutile	TiO_2	4.250	79.90	18.80	6.27	9.4
Kyanite	Al_2SiO_5	3.674	162.05	44.11	5.51	8.8
Corundum	Al_2O_3	3.988	101.96	25.57	5.11	8.5
Stishovite	SiO_2	4.287	60.09	14.02	4.67	7.0

† Data taken from S. P. Clark, Jr. (ed.), Handbook of Physical Constants, sec. 5, *Geol. Soc. Am. Mem.* 97, 1966.
‡ Volume units are cubic centimeters per mole.   To obtain volume per atom pair, multiply volume per atom by 2.

TABLE 2.5

Comparison of Volumes Occupied by Various Atomic Configurations

	Molar Volume, cm³/mole	Volume per Oxygen Atom, cm³/mole
$Al_2O_3 + SiO_2$		
Corundum + coesite	46.21	9.2
Corundum + quartz	48.26	9.7
Corundum + tridymite	52.10	10.4
Kyanite	44.11	8.8
Sillimanite	49.91	10.0
Andalusite	51.54	10.3
$2MgO + SiO_2$		
2 Periclase + coesite	43.14	10.8
2 Periclase + quartz	45.19	11.3
Periclase + enstatite	42.65	10.6
Olivine	43.79	10.9
$MgO + Al_2O_3$		
Periclase + corundum	36.82	9.2
Spinel	39.72	9.9

From our considerations of the variation of bulk modulus with volume and of the volume per oxygen atom for oxides it is evident that most oxide compounds, and especially silicate minerals, exhibit distinctly different behavior from ionic compounds such as the alkali halides. The silicates appear to act like a single material under different states of compression rather than a family of chemical compounds possessing similar ionic structure. This may be due to the close-packed oxygen-oxygen contacts in silicate structures.

PROBLEMS

2.1 A gas has a molecular weight of 44 and molecular radius of 1.6 Å. Its density is 0.61 g/cm³ at a pressure of 225 atm and a temperature of 27°C. Based on the van der Waals theory, what are the critical temperature and pressure for this gas?

2.2 An unknown gas has the following P, V, T relationship for 1 mole of gas. At $P = 10$ atm, $V = 4l$, the temperature is 500.2°K. At $P = 100$ atm, $V = 0.6l$, the temperature is 799.3°K.

(a) Is the gas ideal?

(b) Calculate the van der Waals constants (a and b) for the gas. Are they sensitive to an error of 1 deg in the temperature?

(c) Using Table 2.1, determine the nature of the unknown gas.

2.3 The function $f(V) = (V - V_c)^3$ must become zero at the critical point. Use this fact and the van der Waals equation to find the critical constants in terms of a, b, and R. (*Hint:* If two functions are equal to the same constant each of the power terms must be separately equal. Given $a_1 x^3 + a_2 x^2 + a_3 x = 1$ and $b_1 x^3 + b_2 x^2 + b_3 x = 1$, $a_1 = b_1$, etc.)

2.4 Show that the virial coefficient, B, in the equation $PV = RT + BP$ must be related to the van der Waals coefficients approximately by $B = b - a/RT$.

2.5 A liquid at 300°K has a compressibility of 5×10^{-5} at 1 atm and of 1×10^{-5} at 1000 atm. What will its compressibility be at 3500 atm? What would you predict the coefficient of thermal expansion to be at 1 atm?

2.6 Prove that $dB_T/dP = (n + 2)/3 + \frac{4}{3}$ as $P \to 0$ starting from

$$B_T = \left[P + A \frac{(n - 2)e^2}{(n + 2)x^4} \right] \frac{n + 2}{3}$$

and noting that

$$\frac{1}{V} \frac{dV}{dP} = \frac{3}{x} \frac{dx}{dP}$$

2.7 Periclase (MgO) has a compressibility of 5.97×10^{-7} atm^{-1} at 1 atm, 300°K. Assuming the ideal Pauling relationship, estimate the compressibilities of ZnS (sphalerite, $\rho = 4.088$), PbS (galena, $\rho = 7.597$), CaS (oldhamite, $\rho = 2.602$). Compare with the experimental values of

Galena	1.96×10^{-6} atm^{-1}
Oldhamite	2.32×10^{-6}
Sphalerite	1.30×10^{-6}

REFERENCES

Anderson, O. L., and N. Soga: A Restriction to the Law of Corresponding States, *J. Geophys. Res.*, **72**:5754 (1967).

Castellan, Gilbert W.: "Physical Chemistry," p. 33, Addison-Wesley Publishing Company, Inc., Reading, Mass., 1964.

Moelwyn-Hughes, E. A.: "Physical Chemistry," 2d ed., p. 586, Pergamon Press, New York, 1961.

Pauling, Linus: "The Nature of the Chemical Bond," Cornell University Press, Ithaca, N.Y., 1960.

chapter three **Energy**

The second property of a chemical system which must be known in order to determine its equilibrium state is the energy content of each of the possible arrangements of the atoms and molecules present.   Energy is stored in two forms: kinetic energy associated with the motion of the atoms and potential energy associated with the electrostatic interaction between the atoms.   Although the amount of energy contained by a system is primarily a function of the chemical form and physical state of the atoms present, for any given atomic arrangement the energy content will vary with environmental conditions.   Changes in these conditions which increase the temperature of the substance lead to larger amounts of stored energy.   Changes which increase the volume of the substance also increase the stored energy in most cases.   As no evidence for the net generation or destruction of energy during chemical processes has been found, the first law of thermodynamics applies.   This law states that energy is conserved during all natural processes.   If energy is added to a system during any process it must be balanced by a corresponding loss from the surroundings.

Energy transfer is generally accomplished either by heat flow or by reapportionment of space between the system and the surroundings. This means that the energy content of most substances depends only on temperature and volume. Thus the change in energy content of the substance can be written as follows:

$$dE = \left(\frac{\partial E}{\partial T}\right)_V dT + \left(\frac{\partial E}{\partial V}\right)_T dV$$

ENERGY ASSOCIATED WITH MOLECULAR MOTION

The temperature of any material reflects the degree of agitation of its component atoms. At constant volume, energy can be added only by increasing the kinetic energies of these motions. The energy required to raise the level of thermal agitation in a specified amount of material so that its temperature rises 1 deg is defined as its *heat capacity*. If this heating is carried out at constant volume, the heat capacity is designated C_V. In partial-derivative form, by definition

$$\left(\frac{\partial E}{\partial T}\right)_V = C_V$$

Although molecular motion can consist of translation, rotation, and vibration, only rarely do all three types of motion contribute to the energy of a substance. For example, for gases at low temperatures only translational and rotational modes of motion are important. On the other hand, for solids only vibrational motion is generally possible. The available types of motion interact in such a way that the number of modes of motion available to any given atom is always three. The identity of the modes of motion will change if the physical state of the atom is changed. Thus an Ar atom has three modes of translational motion in the gaseous state but has three modes of vibrational motion when frozen into a solid.

In order to raise the temperature of any substance the energy associated with all available modes of motion must be increased. If the amount of energy required by each mode for this increase were identical, it follows that the heat capacity per atom of all substances would be identical. In other words, the energy required per mole of *atoms* at constant volume to raise by 1 deg the temperature of solid pyrite would be the same as that for liquid water and that for gaseous methane. Also, heat capacity would not change with temperature or pressure.

Although room-temperature heat capacities per atom for a wide range of substances are similar, there are real differences. Also, almost all materials show variations of heat capacity with temperature. These deviations from the very simple model given above stem from two sources. First, twice as much energy is required to raise the temperature of a vibrational mode by 1 deg than for a translational or rotational mode. Second, vibrational modes accept their full share of energy only at relatively high temperatures. They accept negligible energy close to absolute zero. Whereas these exceptions to the simple model permit a complete range of heat capacities from twice the universal value suggested above (for solids at very high temperature) to zero (for solids near absolute zero), in the room-temperature range the two effects partially cancel, leading to values which fall within ±50 percent of this simple predicted heat capacity.

The reason for the twofold larger heat capacities of vibrational modes is related to the nature of the motion involved. For translation and rotation, the energy added goes into kinetic form only. For vibrational motion, the energy added is used to increase both the kinetic and potential energy of the atoms involved. The potential-energy increase arises from the stretching of the electrostatic bonds between atoms during a vibration, and the kinetic-energy increase is associated with the motion itself. At the extremes of the vibratory path the participating atoms are momentarily at rest and all the energy is in the potential form. By contrast, as the atoms pass the vibratory midpoint they are traveling at maximum velocity, all the energy being in kinetic form. The average value of the kinetic-energy contribution is the same as the potential-energy contribution. Regardless of the position of the atoms at any given instant, the sum of the kinetic and potential energy is constant. Since two atoms are involved in opposing motion and each experiences an increase in its energy content because of the motion, twice as much energy must be supplied to the system to produce a given increase in temperature than would be necessary for a single translational or rotational mode.

The second exception to the simple model of heat capacity is the failure of vibrational modes to accept their full component of energy at low temperatures. The problem stems from the fact that the energies for any type of motion are quantized. In other words, there are certain definite energies which a mode of motion can assume. These levels are separated by energy gaps. One requirement for equipartition of energy is that the mean thermal-agitation energy in the medium be comparable to the width of these gaps. If this is not the case, the transitions between permissible levels will be greatly impeded.

This problem is serious only for vibrational levels, since they are generally separated by gaps much larger than the ambient thermal energy of the atoms at room temperature. For translational motion the spacing of levels is so small that free movement between levels is possible even close to absolute zero.   Rotational modes normally have an intermediate spacing such that temperatures of only tens of degrees are necessary before the ambient thermal energy is adequate to move the atoms or molecules freely from state to state.

Since almost all materials of interest show only partial activation of vibrational modes, a few words regarding the manner in which this activation takes place is in order.   The simplest type of vibration is that which takes place between two atoms in a diatomic gas.   Each molecule undergoes an independent vibratory motion along its bond axis.   For any type of vibration the molecular dumbbell has its own characteristic frequency, ν.   This frequency depends on the mass of the atoms and the rigidity of the bond binding them together.   Even at absolute zero the molecule would vibrate with its fundamental frequency.   The energy associated with this ground-state motion has been shown to equal $\frac{1}{2}h\nu$, where h is Planck's constant.   In addition to the ground state, a series of higher energy states of vibration exists.   The energy separation between the lowest levels, $h\nu$, is proportional to the fundamental frequency. Since the difficulty of activation is proportional to the magnitude of the energy gap, higher temperatures are required to populate the elevated vibrational states of a molecule with a high fundamental frequency than for one with a low frequency.

As will be shown in Chap. 4, from a knowledge of the spacing of the lower vibrational states, $h\nu$, and of the mean state of thermal agitation,

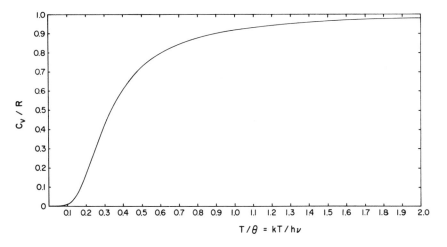

Fig. 3.1. Activation of the heat capacity of a simple diatomic oscillator plotted as a function of $kT/h\nu$. R, the gas constant, is the fully activated value of C_V.

kT, it is possible to predict what fraction of the molecules are in the ground state and in each of the elevated states. Once this is known, the total energy associated with vibration can be calculated at all temperatures. The change in this energy per degree of temperature rise is the heat capacity due to this vibration. As shown in Fig. 3.1, when a plot of vibrational heat capacity for an ideal diatomic molecule is made against the ratio $kT/h\nu$, a curve is obtained which starts at zero at absolute zero, rising to half its fully activated value when $kT/h\nu$ reaches one-third. This is where the mean energy available for exciting motion equals one-third of the spacing between energy levels. Beyond $kT/h\nu$ equals unity, the predicted heat capacity levels off at its equipartition value. This curve is calculated by using the Einstein equation for vibrational heat capacity.

Although any given vibration can be characterized simply by its fundamental frequency, it proves to be more convenient to assign it a characteristic temperature, θ, the temperature at which $kT = h\nu$. The characteristic temperature is then that temperature at which the vibrational heat capacity is 92 percent of the fully "activated" value. One-half activation occurs at the temperature $\theta/3$.

With these basic concepts in mind, it is possible to consider in more detail the heat capacities of various materials.

HEAT CAPACITY OF GASES

As mentioned in Chap. 2, the heat capacity at constant volume for translational motion is $\frac{3}{2}k$. Thus, for each translational mode the molecule (or atom) should absorb $k/2$ units of energy per degree temperature rise. The rotational modes would each have an identical heat capacity and each of the vibrational modes twice this heat capacity. For 1 mole of molecules (or atoms) undergoing a given mode the heat capacity would be 6.02×10^{23} times as large. The result is about 1 cal/deg-mole for translational and rotational modes and 2 cal/deg-mole for vibration.

The only modes of motion available to the atoms in a monatomic gas are translational. Since there are three possible orthogonal directions in which atoms may move, C_V should be 3 cal/deg-mole. Since translational motion is fully activated even close to absolute zero, this heat capacity should remain constant with temperature. Observation confirms this prediction.

Diatomic gases such as H_2 or O_2 have six possible modes of motion per molecule (2 atoms/molecule $\times$ 3 modes/atom). In addition to three translational modes of the molecule as a unit, it can also have two rotational modes. These rotations occur on axes orthogonal to that of the molecular axis.   Rotation on the bond axis is not an available mode because nearly the entire mass of the two atoms (their nuclei) lies precisely on this axis.   The sixth mode is a vibration along the bond axis. The heat capacity corresponding to full activation of these modes is then

$$C_V = 3 \times 1 + 2 \times 1 + 1 \times 2 = 7 \text{ cal/deg-mole}$$

At temperatures at which the activation of the vibrational modes is negligible the heat capacity is 5 cal/deg-mole.   Close to absolute zero, where rotational uptake is impeded, the heat capacity drops to 3 cal/deg-mole.   The temperature dependence of the heat capacity of H_2 and O_2 is shown in Fig. 3.2   The characteristic temperature for their vibrations are, respectively, 6300 and 2280 deg absolute.

For triatomic molecules the theoretical heat capacity depends on whether the molecule is linear (all three atoms in line) or nonlinear. Whereas both types have three translational modes out of nine, three modes of rotation are available to the nonlinear molecule and only two to

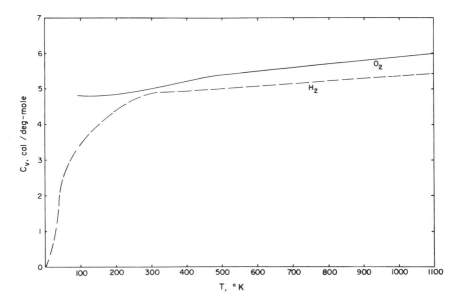

Fig. 3.2. Heat capacity at constant volume for the gases O_2 and H_2. These gases have characteristic Einstein temperatures for vibration of 2280°K for O_2 and 6300°K for H_2.   The translation-plus-rotation contribution is 5 cal/deg-mole, and the ideal fully activated heat capacity is 7 cal/deg-mole for both gases. (*Data from K. K. Kelley, U.S. Bur. Mines Bull.* 584, 1960, *and "Handbook of Chemistry and Physics," 45th ed., The Chemical Rubber Publishing Company, Cleveland, Ohio.*)

the linear.   Thus the linear molecule would have one more vibrational mode than the nonlinear molecule.   The fully activated heat capacities would be

$$C_V = 3 \times 1 + 2 \times 1 + 4 \times 2 = 13 \text{ cal/deg-mole (linear)}$$

and

$$C_V = 3 \times 1 + 3 \times 1 + 3 \times 2 = 12 \text{ cal/deg-mole (nonlinear)}$$

CO_2 and H_2O are examples of triatomic gases.   As shown in Fig. 3.3, their heat capacities are complex functions of temperature because each vibration has its own frequency and hence characteristic temperature. For H_2O, a nonlinear molecule, the three characteristic temperatures are 5370, 5220, and 2280 deg absolute.   In Table 3.1 the heat capacities calculated from these temperatures are compared with the observed values.

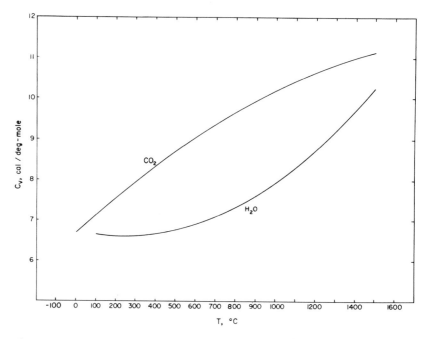

Fig. 3.3. Heat capacity at constant volume for the gases CO_2 and H_2O. The ideal fully activated heat capacity would be 13 cal/deg-mole for CO_2 and 12 cal/deg-mole for H_2O. The rotational-plus-translational contribution is 5 cal/deg-mole for CO_2 and 6 cal/deg-mole for H_2O. *(Data from K. K. Kelley, U.S. Bur. Mines Bull. 584, 1960, and "Handbook of Chemistry and Physics," 45th ed., The Chemical Rubber Publishing Company, Cleveland, Ohio.)*

Since almost all gases with four or more atoms are nonlinear, their heat capacities for full excitation can be computed from the following general relationship:

$$C_V = 3 \times 1 + 3 \times 1 + (3n - 6)2 = 6(n - 1) \qquad \text{cal/deg-mole}$$

For such gases each of the $3n - 6$ vibrational modes would have its own characteristic temperature.

TABLE 3.1

Heat Capacity of Water Vapor†

Temperature, °K	Calculated C_V, cal/deg-mole	Observed C_V, cal/deg-mole
300	6.02	
400	6.18	
500	6.40	6.51
600	6.67	6.73
700	6.95	6.93
800	7.23	7.15
900	7.53	7.38
1000	7.83	7.61

† Data from K. K. Kelley, *U.S. Bur. Mines Bull.* 584, 1960, and "Handbook of Chemistry and Physics," 45th ed., The Chemical Rubber Publishing Company, Cleveland, Ohio.

HEAT CAPACITY FOR SOLIDS AND LIQUIDS

Since the atoms present in solids generally undergo only vibrational motion, their heat capacities, C_V, should approach $3n \times 2$ or $6n$ cal/deg-mole at high temperatures, where n is the number of atoms per formula weight. As absolute zero is approached, the heat capacities approach zero.

The pattern of increase in heat capacity is, in general, considerably more complex than that shown above for a simple diatomic molecule. The C_V versus T curve will be a composite of $3n$ separate activation curves, one for each mode of vibration. Because the vibration frequencies associated with the various modes are, in general, not all the same, the composite curve will not follow that of a single Einstein function. Further, only $3(n - 1)$ of the vibrational modes can be characterized by a single frequency. The other three consist of a whole spectrum of frequencies and hence are not subject to the Einstein treatment.

Debye showed that the frequency spectrum could be integrated to yield a heat capacity versus temperature relationship which could be written in terms of a characteristic temperature, θ_D, in the same manner as the Einstein function is written in terms of θ_E. The so-called Debye temperature is equal to $h\nu_{max}/k$, where ν_{max} is the maximum frequency in the spectrum. Plots of heat capacity versus fraction of Einstein and fraction of Debye temperature are given in Fig. 3.4. The Debye vibrations show a higher degree of activation for any given fraction of the characteristic temperature. This is because the mean frequency for a Debye mode is considerably less than the maximum frequency used to define the Debye temperature.

For most solids a good fit to the observed heat capacity can be obtained by summing a limited number of Debye and Einstein components. Although the $\theta_D s$ and $\theta_E s$ obtained by these empirical separations have no theoretical significance, they provide a useful means of expressing the temperature dependence of heat capacity. For example, the heat capacity of SrO can be written

$$C_V = 6f_D\left(\frac{261}{T}\right) + 6f_E\left(\frac{444}{T}\right)$$

The characteristic temperature for the Debye mode is 261°K and that

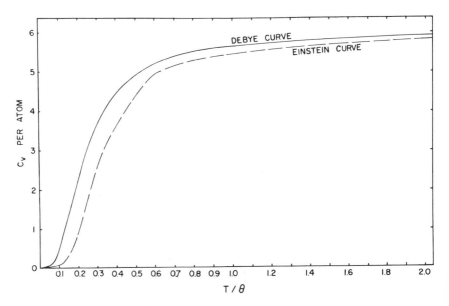

Fig. 3.4. Comparison of Debye and Einstein heat-capacity activation for solids.

TABLE 3.2

**Heat Capacity per Atom
for Representative Solids†**

Compound	Heat Capacity per Atom at 298°K, cal/deg-mole
Sylvite (KCl)	6.10
Sphalerite (ZnS)	5.50
Fluorite (CaF₂)	5.34
Calcium oxide (CaO)	5.12
α Quartz (SiO₂)	3.54
Forsterite (Mg₂SiO₄)	4.02
Periclase (MgO)	4.52
Corundum (Al₂O₃)	3.78
Silicon carbide (SiC)	3.21
Diamond (C)	1.45

† Heat-capacity data from K. K. Kelley, *U.S. Bur. Mines Bull.* 584, 1960.

for the Einstein mode, 444°K. The separate curves and their sum are shown in Fig. 3.5.

The heat capacities on a per atom basis at room temperature for a number of solids of geologic interest are given in Table 3.2. It is clear

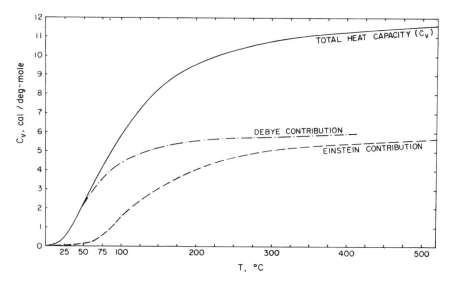

Fig. 3.5. Heat capacity for SrO as a function of temperature. The total C_V may be expressed as the sum of a Debye mode contribution with $\theta_D = 261°K$ and an Einstein mode contribution with $\theta_E = 444°K$. Data from K. K. Kelley, *U.S. Bur. Mines Bull.* 584, 1970.

that there is a considerable variation in the degree of activation of the vibrational modes at room temperature. This range stems from a difference in the bond strengths for various solids; the more rigid the bonding, the lower is the room-temperature heat capacity. For molecules with strong covalent bonds, such as Al_2O_3, the heat capacity is fairly low. For solids such as metallic iron, the intermolecular forces are relatively weak with a commensurately high heat capacity.

The room-temperature heat capacities of the oxides show a particularly interesting property. They are additive. For example, the molar heat capacity of sillimanite, Al_2SiO_5, is equal to the sum of that for quartz, SiO_2, and corundum, Al_2O_3. This suggests that, in oxides, the vibration frequencies which are not fully activated at room temperature are largely dependent on the nature of the individual cation-oxygen bonds, and not on the atomic arrangement in the complex solid. On the other hand, the heat capacities at 50°K are not additive, implying that the important vibrations in this region are strongly dependent on the particular atomic structure.

A list of heat capacities for simple oxides at room temperature is given in Table 3.3. With this list, a very good estimate of the heat capacity of any complex oxide can be made. Some examples are given in Table 3.4. This table also illustrates the nonadditivity of oxide heat capacities at 50°K.

The liquid of prime importance in natural systems is water. The heat capacity of water (18 cal/deg-mole) is surprisingly large compared with that of ice (~9 cal/deg-mole) and with that of water vapor (~8 cal/deg-mole). The value for liquid water corresponds to the theoretical heat capacity for a solid with all its vibrational modes fully activated.

ENERGY ASSOCIATED WITH VOLUME OCCUPIED

In addition to taking up energy by increased molecular agitation, all real substances take up energy when they expand (except under conditions of extreme pressure). This energy goes into stretching electrostatic bonds. If the change in energy content is considered as a function of volume at constant temperature, only this electrostatic component is involved.

The amount of energy required to cause expansion should then be the internal pressure operating in the substance times the change in volume. Hence

$$dE = P_{int}\, dV = (P_{thermal} - P_{ext})\, dV$$

The value of the thermal pressure for any van der Waals substance

TABLE 3.3

**Heat Capacities of the Simple Oxides
at Room Temperature**

Oxide	C_P†	n	C_P/n
B_2O_3	15.00	5	3.00
BeO	6.06	2	3.03
CO_2‡	9.34	3	3.11
H_2O‡	9.60	3	3.20
SO_3‡	13.59	4	3.40
SiO_2	10.62	3	3.54
Al_2O_3	18.88	5	3.78
TiO_2	13.16	3	4.40
Li_2O	13.38	3	4.46
ZrO_2	13.42	3	4.47
MgO	9.03	2	4.52
ZnO	9.62	2	4.81
ThO_2	14.76	3	4.92
Fe_2O_3	24.81	5	4.96
CaO	10.23	2	5.12
MnO	10.54	2	5.27
FeO	10.56	2	5.28
NiO	10.60	2	5.30
SnO	10.60	2	5.30
SrO	10.76	2	5.38
BaO	10.95	2	5.67
Na_2O‡	16.84	3	5.62
K_2O‡	17.50	3	5.82

† Heat-capacity data from K. K. Kelley, *U.S. Bur. Mines Bull.* 584, 1960. Units are calories per degree-mole.
‡ Approximate values, calculated indirectly; for example, $CO_2 = CaCO_3 - CaO$.

has been shown to equal a function of the fraction of available volume divided by the compressibility. Hence

$$\beta = \frac{X}{P_{\text{thermal}}} \quad \text{and} \quad P_{\text{thermal}} = \frac{X}{\beta}$$

Since the coefficient of thermal expansion for such substances is $\alpha = X/T$, the thermal pressure is equal to the absolute temperature times the ratio of the coefficient of thermal expansion to the compressibility. It follows that

$$dE = \left(T\frac{\alpha}{\beta} - P\right) dV$$

TABLE 3.4

**Comparison of Heat Capacities Obtained
by Summation with Measured Values†
(Units Are Calories per Degree-mole)**

Temp., °K					Predicted Observed
	Quartz + corundum	$= \Sigma$		kyanite	
	$SiO_2 + Al_2O_3$	$= \Sigma$		Al_2SiO_5	
300	$10.62 + 18.88$	$= 29.50$		29.10	1.01
50	$1.38 + 0.36$	$= 1.74$		0.84	2.07
	Cristobalite + corundum	$= \Sigma$		sillimanite	
	$SiO_2 + Al_2O_3$	$= \Sigma$		Al_2SiO_5	
300	$10.56 + 18.88$	$= 29.44$		29.31	1.005
50	$1.56 + 0.36$	$= 1.92$		1.79	1.07
	Periclase + corundum	$= \Sigma$		spinel	
	$MgO + Al_2O_3$	$= \Sigma$		$MgAl_2O_3$	
300	$9.03 + 18.88$	$= 27.91$		27.71	1.006
50	$0.22 + 0.36$	$= 0.58$		0.88	0.66
	Periclase + quartz	$= \Sigma$		pyroxene	
	$MgO + SiO_2$	$= \Sigma$		$MgSiO_3$	
300	$9.03 + 10.62$	$= 19.65$		19.62	1.001
50	$0.22 + 1.38$	$= 1.60$		1.33	1.20
	Wollastonite + rutile	$= \Sigma$		sphene	
	$CaSiO_3 + TiO_2$	$= \Sigma$		$CaTiSiO_5$	
300	$20.38 + 13.16$	$= 33.54$		33.21	1.01
50	$2.70 + 1.41$	$= 3.71$		3.57	1.04
	Anhydrite + ice	$= \Sigma$		gypsum	
	$CaSO_4 + 2H_2O$	$= \Sigma$		$CaSO_4 \cdot 2H_2O$	
300	$23.82 + (2)(9.30)$	$= 42.42$		44.46	0.955
50	$3.86 + (2)(1.90)$	$= 7.66$		7.60	1.01
	Periclase + ice	$= \Sigma$		brucite	
	$MgO + H_2O$	$= \Sigma$		$Mg(OH)_2$	
300	$9.03 + 9.30$	$= 18.33$		18.43	0.994
50	$0.22 + 1.90$	$= 2.12$		1.18	1.80

† Heat-capacity data from K. K. Kelley, *U.S. Bur. Mines Bull.* 584, 1960.

As shown in Chap. 2, α for an ideal gas is $1/T$ and β is $1/P$. The thermal pressure, $T(\alpha/\beta)$, is then equal to P, the external pressure, and $(\partial E/\partial V)_T$ is zero. This is to be expected since by definition an ideal gas is one in which no electrostatic interactions take place. The only way the energy content of an ideal gas can be changed is to change its degree of molecular agitation. The internal energy of an ideal gas is independent of its molar volume.

Although it cannot be proved here, the relationship $P_{thermal} =$

$T(\alpha/\beta)$ is universally valid. It applies to solids as well as to van der Waals substances.

For an ideal ionic solid the thermally induced vibrations maintain separations between the atoms exceeding those predicted by a simple balance between the forces of attraction and repulsion. If, for example, the mean separation distance exceeds the "equilibrium" distance, x_{equil}, by an amount Δx, then a *thermal force*, F_{thermal}, must balance the excess of the electrostatic attraction force plus external force over the force of repulsion. For a single ionic pair we may approximate the balance of forces by

$$F_{\text{thermal}} = \frac{e^2}{(x_{\text{equil}} + \Delta x)^2} - \frac{B}{(x_{\text{equil}} + \Delta x)^n} + F_{\text{ext}}$$

Since

$$B = x_{\text{equil}}^{n-2} e^2$$

$$F_{\text{thermal}} = \frac{e^2}{(x_{\text{equil}} + \Delta x)^2} \left[1 - \left(\frac{x_{\text{equil}}}{x_{\text{equil}} + \Delta x} \right)^{n-2} \right] + F_{\text{ext}}$$

which for $\Delta x \ll x_{\text{equil}}$ can be approximated by

$$F_{\text{thermal}} = (n - 2) \frac{e^2}{x_{\text{equil}}^2} \frac{\Delta x}{x_{\text{equil}}} + F_{\text{ext}}$$

$$P_{\text{thermal}} \cong (n - 2) \frac{e^2}{x_{\text{equil}}^4} \frac{\Delta x}{x_{\text{equil}}} + P_{\text{ext}}$$

For a crystal such as KCl (following the treatment in Chap. 2),

$$P_{\text{thermal}} = \frac{(n - 2)1.7}{3} \frac{e^2}{x_{\text{equil}}^4} \frac{\Delta V}{V_{\text{equil}}} + P_{\text{ext}}$$

Since for KCl the term $[(n - 2)1.7/3](e^2/x_{\text{equil}}^4)$ is equal to 1×10^6 atm, and at room temperature $\Delta V/V_0$ (the volume change from absolute zero to T) is about 2×10^{-2}, the thermal pressure is about 2×10^4 atm. This result can be compared with that obtained from the ratio of the coefficient of thermal expansion for KCl, 1.1×10^{-4} deg^{-1} divided by its compressibility 5×10^{-6} atm^{-1} times the absolute temperature, 300 deg [hence $T(\alpha/\beta)$]. The result of 6600 atm is the same order of magnitude as that obtained from the simple ionic-bonding model.

In summary, the energy content of any substance can be raised either by increasing its degree of molecular agitation (hence its temperature) or by increasing the separation between its atomic or molecular units (hence its volume).   The total internal-energy change is given by

$$dE = C_V\, dT + \left(T\frac{\alpha}{\beta} - P\right) dV$$

This result is completely general.

VARIATION OF STORED ENERGY WITH TEMPERATURE AT CONSTANT PRESSURE

For most natural processes, temperature and volume change together. Thus the contribution of changes of thermal agitation $(C_V\, dT)$ and of interaction energy $[(\alpha T/\beta - P)\, dV]$ must be simultaneously considered. Since many natural processes are isobaric (occur at constant pressure) the total energy change under these conditions is of interest.   Since for isobaric processes $\Delta V = V\alpha\, \Delta T$, it follows that

$$dE = \left(C_V + \frac{TV\alpha^2}{\beta} - PV\alpha\right) dT$$

where the first term represents the contribution of molecular motion, and the second and third that of bond stretching.

If the sources of the energy involved in an isobaric heating are considered, rather than the mode of storage of the energy, it is convenient to combine the first two instead of the last two terms.   The sum $C_V + TV\alpha^2/\beta$ represents the amount of heat which must be added to the substance at constant pressure to raise its temperature 1 deg.   It is the sum of the energy, C_V, required to increase the degree of thermal agitation, and the energy, $TV\alpha^2/\beta$, required to expand the solid in response to the thermal pressure.   This sum is defined as the heat capacity at constant pressure, C_P.

The increase in energy content of the solid, dE, is always somewhat less than the heat added because some of the heat, $PV\alpha\, dT$, is used in pushing back the surroundings.   In terms of C_P, the internal-energy increase at constant pressure is

$$dE = (C_P - PV\alpha)\, dT$$

or, in partial-derivative form,

$$\left(\frac{\partial E}{\partial T}\right)_P = C_P - PV\alpha$$

Since $PV \alpha \, dT = P \, dV$, at constant pressure

$$dE = C_P \, dT - P \, dV$$

As will be shown in the next chapter, the amount of heat transferred between the system and the surroundings during isobaric processes has special importance in dealing with entropy changes. For this reason a composite property of the system equal to the sum of the internal energy, E, and the product, PV, is defined as the *enthalpy*, H, of the system. Hence

$$H = E + PV$$

and

$$dH = dE + P \, dV + V \, dP$$

Since for isobaric processes $dP = 0$,

$$dH = dE + P \, dV$$

and thus at constant pressure

$$dH = C_P \, dT$$

In this way the amount of heat transferred between the system and the surroundings, $C_P \, dT$, can be related directly to the change in the enthalpy of the system.

It is of interest to consider the relative magnitudes of the amount of heat required to increase the molecular motion and that required to expand the substance against the total confining pressure ($P_{int} + P_{ext}$). The ratio of these two energies is equal to $(C_P - C_V)/C_V$. If C_P/C_V is defined as γ, then

$$\frac{C_P - C_V}{C_V} = \gamma - 1 = \frac{TV\alpha^2}{\beta C_V}$$

For an ideal gas, β and α are $1/P$ and $1/T$, respectively. Thus

$$\frac{C_P - C_V}{C_V} = \frac{R}{C_V}$$

The amount of energy per degree temperature increase required to expand an ideal gas during an isobaric heating is thus 2 cal/deg-mole. This can be compared with the C_V of various gases at room temperature: Ar, ~ 3

cal/deg-mole; O_2, $\sim$5 cal/deg-mole; and H_2O, $\sim$6 cal/deg-mole. The expansion energy makes up 25 to 40 percent of the total energy added.

For a van der Waals substance,

$$\alpha = \frac{1}{V}\frac{1}{(\partial T/\partial V)_P} = \frac{1}{V(P/R + 2ab/RV^3 - a/RV^2)}$$

Neglecting the $2ab/RV^3$ term, which is very small compared with the other two,

$$V\alpha = \frac{R}{P - a/V^2}$$

Also, $T\alpha/\beta$ for a van der Waals substance is P_{thermal} which is equal to $P + a/V^2$. Hence

$$\frac{C_P - C_V}{C_V} = \frac{TV\alpha^2}{\beta C_V} = \frac{1}{C_V}\frac{T\alpha}{\beta}V\alpha$$

$$\cong \frac{1}{C_V}\frac{R}{P - a/V^2}\left(P + \frac{a}{V^2}\right)$$

$$= \frac{R}{C_V}\frac{P + a/V^2}{P - a/V^2}$$

Water vapor at 127°C (400°K) provides a good example of the influence of molecular attraction. Were this attraction negligible, $\gamma - 1$ would equal R/C_V, or 0.305. The van der Waals constant, a, for water is 5.46 liters²-atm/mole². Thus, at 1 atm, a/V^2 is equal to 5×10^{-3}. This correction raises $\gamma - 1$ to 0.308 (hence by 1 percent). The magnitude of this correction would increase with rising pressure.

Whereas for gases the internal pressure is very small, for liquids and solids the internal pressures are of the order of 10^4 atm. At ordinary pressures the work expended by the solid in pushing back its surroundings is negligible compared with the energy required to pull apart the atoms in the solid itself. Thus

$$\left(\frac{\partial E}{\partial V}\right)_T \cong \frac{T\alpha}{\beta}$$

For isobaric heating the magnitude of the volume change is so very small that the amount of the energy required for the expansion is small compared with that for gases. Thus $(C_P - C_V)/C_V$ is much smaller for solids than for gases.

Although a general relationship between α and β is not available for solids, empirical observation demonstrates that the ratio $V\alpha^2/\beta C_P^2$ remains nearly constant with temperature and pressure for a given solid. That this constancy is reasonable can be shown as follows: As mentioned previously, both the heat capacity and coefficient of thermal expansion depend upon the degree to which the vibrations have been activated. Consequently, α is roughly proportional to C_P. Both are zero at absolute zero and rise toward nearly constant values at high temperature. Volume and compressibility do not change significantly with temperature.

$$C_P - C_V = \frac{TV\alpha^2}{\beta} = \frac{V\alpha^2}{\beta C_P^2} C_P^2 T = A C_P^2 T$$

Hence

$$\frac{C_P - C_V}{C_V} = \frac{A C_P^2 T}{C_V} = \frac{A C_P^2 T}{C_P - A C_P^2 T} = \frac{A C_P T}{1 - A C_P T} = \gamma - 1$$

The relationship between γ and T is written in terms of C_P rather than C_V, since for solids C_P alone can be determined by experiment. Whereas for a gas the volume can be held constant during heating, this is not practical when dealing with solids. If C_V is needed, it must be obtained from a combination of theory and experiment from the equation

$$C_V = C_P - \frac{TV\alpha^2}{\beta}$$

or

$$C_V = C_P - A C_P^2 T$$

The values of C_V, C_P, α, β, A, $C_P - C_V$, and γ are given in Table 3.5 for copper at several different temperatures.

For water at room temperature the energy associated with overcoming the internal pressure during an isobaric heating, $TV\alpha^2/\beta$, is 0.18 cal/deg-mole compared with 18 cal/deg-mole for the energy required to excite molecular motion. Thus $(C_P - C_V)/C_V$ for water is only 0.01.

ADIABATIC CHANGES

Processes carried out in systems thermally isolated from their surroundings are also of importance in the earth sciences. For adiabatic expan-

TABLE 3.5

Thermal Properties of Copper†

T, °K	C_P, cal/ deg- mole	C_V, cal/ deg- mole	α, 10^{-5} deg^{-1}	β, 10^{-6} atm^{-1}	V, cm^3/ mole	$C_P - C_V$, cal/ deg-mole	A, 10^{-5} mole/ cal	γ
50	1.50	1.50	11.4	0.722	7.00	0.00152	1.35	1.00
100	3.85	3.83	31.5	0.731	7.01	0.0230	1.55	1.00
150	4.90	4.86	40.7	0.744	7.02	0.0567	1.57	1.01
200	5.45	5.36	45.3	0.759	7.03	0.0920	1.55	1.02
250	5.74	5.62	48.3	0.773	7.04	0.128	1.56	1.02
300	5.86	5.69	50.4	0.788	7.06	0.165	1.60	1.03
500	6.17	5.86	54.9	0.850	7.12	0.305	1.60	1.05
800	6.63	6.08	60.0	0.935	7.26	0.541	1.54	1.09
1200	7.22	6.22	70.2	1.045	7.45	1.047	1.67	1.16

† After M. W. Zemansky, "Heat and Thermodynamics," 5th ed., McGraw-Hill Book Company, New York, 1968.

sion or compression, changes in internal energy can be accomplished only through work done on the surroundings. Hence

$$dE = -P \, dV$$

An internal exchange of energy between potential and kinetic forms also takes place. For example, during an adiabatic compression, potential energy is released as the atoms move closer together. This energy excites the molecular motions to higher thermal levels, raising the temperature of the substance. As shown above, for any process

$$dE = C_V \, dT + \left(T \frac{\alpha}{\beta} - P \right) dV$$

Since $dE = -P \, dV$ for adiabatic processes,

$$C_V \, dT + T \frac{\alpha}{\beta} \, dV = 0$$

However, since

$$dV = \alpha V \, dT - \beta V \, dP$$

the following relationship results:

$$\left(C_V + \frac{TV\alpha^2}{\beta} \right) dT = TV\alpha \, dP$$

or

$$C_P \, dT = TV\alpha \, dP$$

In partial-derivative form

$$\left(\frac{\partial T}{\partial P}\right)_S = \frac{TV\alpha}{C_P}$$

where S indicates entropy.   As will be shown in the next chapter, adiabatic reversible processes are processes carried out at constant entropy.
For an ideal gas, $\alpha = 1/T$; hence

$$\left(\frac{\partial T}{\partial P}\right)_S = \frac{V}{C_P} = \frac{RT}{C_P P}$$

Nitrogen gas at room temperature has a C_P of ~ 7 cal/deg-mole.   Thus at 1 atm and 25°C the adiabatic gradient for N is

$$\left(\frac{\partial T}{\partial P}\right)_S = \frac{2 \times 300}{7 \times 1} = 85 \text{ deg/atm}$$

Rising air masses whose moisture content remains constant should cool initially at the rate of 8.5 deg/km, because atmospheric pressure drops an average of 0.1 atm/km for heights up to a few kilometers.
    For water, $\alpha = 2.1 \times 10^{-4}$ deg^{-1}, $V = 1.8 \times 10^{-2}$ liter/mole, and $C_P = 18$ cal/deg-mole at 300°K.   The adiabatic gradient for water is thus

$$\left(\frac{\partial T}{\partial P}\right)_S = \frac{(3.00 \times 10^2)(1.8 \times 10^{-2})(2.1 \times 10^{-4})}{1.8 \times 10}$$

$$= 6.3 \times 10^{-5} \text{ deg/cal-liter}$$

$$= 1.5 \times 10^{-3} \text{ deg/atm}$$

In descending from the surface to a depth of 3 km a mass of seawater would be warmed by 0.45 deg (pressure increases 100 atm for each kilometer depth).
    For the mineral periclase at a temperature of 1000°C and a pressure of 1 atm, $\alpha = 4.2 \times 10^{-5}$ deg^{-1}, $V = 1.096 \times 10^{-2}$ liter/mole, and $C_P = 12.7$ cal/deg-mole.   The adiabatic gradient would be

$$\left(\frac{\partial T}{\partial P}\right)_S = 8.8 \times 10^{-4} \text{ deg/atm}$$

If, as some geophysicists believe, the earth's mantle undergoes convective overturn which is rapid with respect to the flow of heat, then the temperature gradient should be that for adiabatic compression.   When the result for periclase is used, a gradient of 300 deg per 1000 km is obtained (pressure in the mantle rises about 330 atm/km).

ISOTHERMAL CHANGES

One other type of process needs consideration: pressure changes which occur at constant temperature.   Although such changes rarely take place in natural systems, such paths are useful in calculating energy changes. For example, if the enthalpy of quartz at 500°C and 400 atm were needed, the most convenient way to calculate it from the enthalpy at standard conditions (that is, 1 atm, 25°C) would be as follows: The change for an isobaric heating from 1 atm, 25°C, to 1 atm, 500°C would first be calculated.   Next the change for an isothermal-pressure increase from 1 atm, 500°C to 400 atm, 500°C would be computed.   By adding the sum of these changes to the enthalpy at standard conditions, the enthalpy at 400 atm, 500°C would be obtained.

We have already shown how the energy of a substance changes for an isobaric-temperature increase, but we have not done this for an isothermal-pressure increase.

As shown above, for all processes

$$dE = C_V\,dT + \left(T\frac{\alpha}{\beta} - P\right)dV$$

For the isothermal-pressure increase the first term is zero and $dV = -V\beta\,dP$.   Hence
$$dE = V(\beta P - \alpha T)\,dP$$

Since $dH = dE + P\,dV + V\,dP$, we have

$$dH = (PV\beta - TV\alpha - PV\beta + V)\,dP$$
or
$$dH = V(1 - \alpha T)\,dP$$

In any process, if the changes in pressure and temperature are very large it is necessary to use a slightly different form of the equations given above.   If changes are large, one must consider α, β, and V to be functions of temperature and pressure.   In these cases, the differential form of any equation must be integrated, taking into consideration the temperature and pressure dependences of all parameters.   For example, for moderate changes in pressure, we may use

$$dH = V(1 - \alpha T)\ dP$$

But for large changes, say, over 1000 atm, the expression must be changed to

$$\Delta H = \int_{P_1}^{P_2} V(1 - \alpha T)\ dP$$

where V and α must be given as explicit functions of pressure.

ENERGY CHANGES ASSOCIATED WITH CHANGES IN ATOMIC ARRANGEMENT

Up to this point the discussion has revolved around how the energy of one particular atomic arrangement varies with temperature and pressure. Our ultimate concern, however, is with the energy differences between different arrangements under one set of conditions.   Although the major contribution to this difference comes from the potential-energy change stemming from the rearrangement itself, the contribution of volume differences and differences in heat capacity are of great importance to our study of chemical equilibria.   If the rearrangement energy were always dominant, for a given set of constituents only one stable configuration (i.e., mineral assemblage) would exist, regardless of the values of temperature and pressure.   Our interest in the subject stems from the fact that for most systems several configurations are possible, each having its own stability field on a pressure-temperature diagram.   This is the case partly because energy differences between two configurations change significantly with temperature and pressure.

An example of the variation of the energy difference between two atomic configurations with environmental conditions is provided by the two $CaCO_3$ polymorphs, calcite and aragonite.   Their thermodynamic properties at 1 atm pressure and room temperature are given in Table 3.6. Calcite has an enthalpy 48 cal/mole greater than that for aragonite.   As $P\,\Delta V_{\text{arag-calc}}$ is only 0.07 cal/mole, ΔE very nearly equals ΔH under these conditions.

If the temperature is raised from 25 to 225°C (pressure held at 1 atm) we can calculate the changes in H and E for both polymorphs, using the relationships

$$\Delta E = \int_{T_1}^{T_2} (C_P - PV\alpha)\ dT$$

$$\Delta H = \int_{T_1}^{T_2} C_P\ dT$$

TABLE 3.6

Thermodynamic Data for Calcite and Aragonite†

	Calcite	Aragonite
Molar volume, cm³	36.94	34.16
Coefficient of thermal expansion, deg⁻¹	1.58×10^{-5}	5.55×10^{-5}
Compressibility, atm⁻¹	1.37×10^{-6}	1.55×10^{-6}
Enthalpy of formation at 298°K, cal/mole	$-288,086$	$-288,134$
Heat capacity, cal/deg-mole	$24.98 + (5.24 \times 10^{-3}T)$ $- (6.20 \times 10^5 T^{-2})$	$20.13 + (10.24 \times 10^{-3}T)$ $- (3.34 \times 10^5 T^{-2})$

† Heat-capacity data from K. K. Kelley, *U.S. Bur. Mines Bull.* 584, 1960. Other data taken from S. P. Clark, Jr. (ed.), Handbook of Physical Constants, *Geol. Soc. Am. Mem.* 97, 1966.

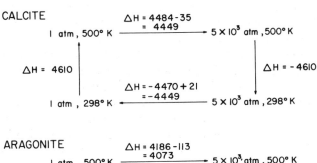

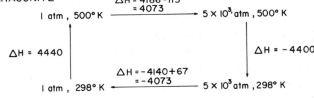

Fig. 3.6. Enthalpy cycle for calcite and aragonite. See text for detailed discussion.

As the $PV\alpha$ term is negligible at 1 atm, E and H both change by the same amount.   As shown in Figs. 3.6 and 3.7, calcite shows an internal energy and enthalpy increase of 4610 cal and aragonite of 4440. ΔE and ΔH both rise from 48 to 218 cal/mole.

If the pressure is now increased to 5000 atm (T held at 225°C) the changes in H and E can be calculated from the following relationships:

$$\Delta E = \int_{P_1}^{P_2} V(\beta P - \alpha T)\, dP$$

and

$$\Delta H = \int_{P_1}^{P_2} V(1 - \alpha T)\, dP$$

where V is the volume at 225°C.

As seen from Fig. 3.7, the internal energies of calcite and aragonite fall 35 and 113 cal/mole, respectively (because of the release of electrostatic energy during compression).   The larger decrease for aragonite

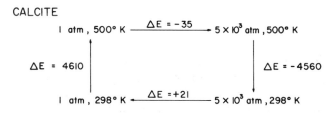

CALCITE

ARAGONITE

CALCITE - ARAGONITE

Fig. 3.7. Energy cycle for calcite and aragonite.   See text for detailed discussion.

reflects its higher coefficient of thermal expansion. This increases the internal-energy difference between the two polymorphs to 296 cal/mole.

The pressure increase results in increases in both the enthalpy of calcite and that of aragonite. Because its molar volume is greater, the increase for calcite exceeds that for aragonite.

If the crystals are returned to the starting conditions first by lowering the temperature to 25°C (at 5000 atm) and then lowering the pressure, the enthalpies of the two polymorphs change by the same amounts as during the heating and compression (see Fig. 3.6). The internal-energy change during cooling does not match that during heating nor does the internal-energy change during decompression balance that for compression. The reasons are as follows: (1) Cooling at high pressure leads to a significant contribution of the $PV\alpha$ term in the ΔE equation; (2) decompression at low temperature reduces the magnitude of the $TV\alpha$ term in the ΔE equation. The two effects cancel one another so that the energy change around the cycle is zero, as it must be.

PROBLEMS

3.1 A hypothetical ideal gas has a heat capacity of 0.19 cal/g-deg at $T = 1°K$, 0.27 cal/g-deg at $T = 100°K$, and 0.59 cal/g-deg at $T = 1000°K$. What is the molecular weight of the gas? Is it linear or nonlinear? How many atoms are there per molecule? Which heat capacity was measured?

3.2 A van der Waals gas has a critical temperature of 400°K. Its molecules have a diameter of 2×10^{-8} cm. At $T = 300°K$, $P = 10$ atm, what will be the difference between the heat capacities, $C_P - C_V$?

3.3 A solid with six atoms per formula and a formula weight of 150 g has a density of 3.00 g/cm³. The Debye temperature is 150°K. If the density decreases 100 ppm for each degree temperature rise and increases 2 ppm for each atmosphere-pressure increase, how much heat is required to raise 1 g of this substance from 200 to 300°C?

3.4 A solid has a heat capacity of 3.5 cal/deg-atom at 300°K and 0.4 cal/deg-atom at 50°K. If half of its vibrations can be approximated by a single Debye function and half by a single Einstein function, what are the approximate Debye and Einstein temperatures?

3.5 What is γ for a solid which at 400°K has an adiabatic gradient of 1×10^{-3} deg/atm and a thermal pressure of 2×10^4 atm?

3.6 The heat capacity (C_P) of sylvite at 600°K is 13.32 cal/deg-mole. The vibrational modes are all fully activated at 600°K. The volume of sylvite at this temperature is 28.28 cm³; the thermal pressure is 1.62×10^4 atm. Find the coefficient of thermal expansion and the compressibility.

3.7 Lava lakes generally have a temperature of about 1100°C. Assuming that the magma travels from its place of origin quickly enough so that negligible heat is lost to wall rocks, calculate the temperature of the magma at a depth of 40 km. The density of basaltic liquid at 1200°C is 2.61; the coefficient of thermal expansion is about 1×10^{-4}. Assume a heat capacity of 0.2 cal/g-deg.

3.8 Calculate the internal-energy and enthalpy changes when water vapor at 10 atm, 400°C, cools to 1 atm, 100°C; condenses; and further cools to 25°C. The required thermodynamic functions are C_P (water vapor) = $7.30 + (2.46 \times 10^{-3}T)$ cal/deg-mole, C_P (liquid H_2O) = 18 cal/deg-mole, $\alpha = 2.1 \times 10^{-4}$ deg^{-1}, $\beta = 0.137/(P + 2800)$ atm^{-1}. Assume ideal gas behavior for water vapor.

chapter four Randomness and Entropy

From the volume and energy content of a chemical system it is possible to determine the degree of randomness of the atoms present.   By randomness we mean the number of different ways in which the atoms can arrange themselves with respect to their geometric locations and their energy content.   The degree of randomness depends heavily on the physical state of the system.   The molecules in a gas can assume a vast number of "visibly" different locations and have a far higher degree of geometric disorder than when frozen onto the ordered lattice points of a solid.   The rise in the number of energy levels available to individual atoms as a substance is heated is a less obvious, but equally important, source of increased randomness.   The greater the amount of energy available to the atoms, the greater is the number of ways it can be distributed among the atoms, and the greater the randomness of the system.

Quantitatively, the randomness of any system can be measured by the number of complexions, W, available to the atoms.   An example will best illustrate how this number can be obtained.   Consider a row of six clear plastic straws, as shown in Fig. 4.1.   If four ball bearings are randomly distributed, one to a straw, there will be 15 "visibly" different

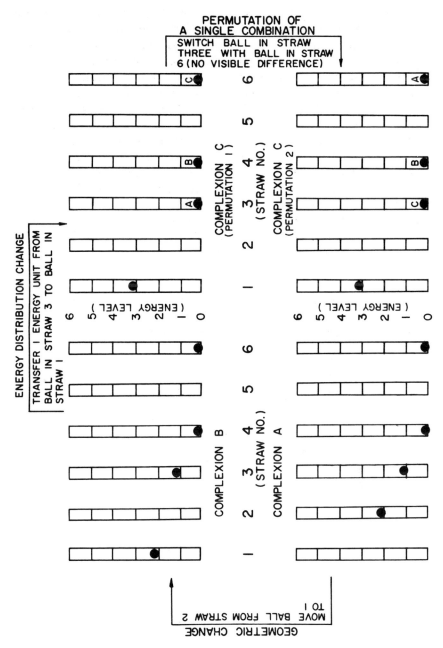

Fig. 4.1. Diagrammatic representation of geometrical randomness, and randomness associated with energy distributions.

arrangements. Note that those arrangements generated by interchanging balls are not counted. This calculation is carried out as follows.

Beginning with the first ball, there are six possible straws in which it could be placed. For the next ball there are five straws in which it could be placed; for the next, four; and the last, three. This gives $6 \times 5 \times 4 \times 3 = 360$ possible arrangements. However, this presumes that we can distinguish each individual ball. The total number of *distinguishable* arrangements is far lower because of the fact that the balls are identical. For each distinguishable arrangement there are four balls to choose from for the first "filled" straw, three for the second, two for the third, and one for the fourth. Hence the total number of arrangements must be divided by $4 \times 3 \times 2 \times 1$, or 24. The true number of complexions of the system is then 360/24, or 15. These arrangements are shown in Table 4.1. If a seventh straw were added, the number of arrangement complexions would rise to 35, that is, $(7 \times 6 \times 5 \times 4)/(4 \times 3 \times 2 \times 1)$. Randomness increases with the number of sites available and hence with *volume*.

If, in addition, each ball has a series of discrete gravitational energy levels represented as heights in the straw, there will be numerous ways in which the available energy can be distributed among the balls. Assume that the levels are equally spaced and that the separation between levels

TABLE 4.1

**Geometric Complexions
(Six Straws; Four Balls)**

	Straw Number					
Arrangement number	1	2	3	4	5	6
1	$\times$	$\times$	$\times$	$\times$	0	0
2	$\times$	$\times$	$\times$	0	$\times$	0
3	$\times$	$\times$	0	$\times$	$\times$	0
4	$\times$	0	$\times$	$\times$	$\times$	0
5	0	$\times$	$\times$	$\times$	$\times$	0
6	$\times$	$\times$	$\times$	0	0	$\times$
7	$\times$	$\times$	0	$\times$	0	$\times$
8	$\times$	0	$\times$	$\times$	0	$\times$
9	0	$\times$	$\times$	$\times$	0	$\times$
10	$\times$	$\times$	0	0	$\times$	$\times$
11	$\times$	0	$\times$	0	$\times$	$\times$
12	0	$\times$	$\times$	0	$\times$	$\times$
13	$\times$	0	0	$\times$	$\times$	$\times$
14	0	$\times$	0	$\times$	$\times$	$\times$
15	0	0	$\times$	$\times$	$\times$	$\times$

is one energy unit.   If the total energy content of the system is three
energy units and if the likelihood of occupation is identical for all levels,
there are 20 ways of distributing the energy for each geometrical arrangement.   This result could be computed as follows.

There are three basic ways in which the energy could be distributed.
It could all be given to one ball.   Since there are four balls, this mode of
distribution gives rise to four complexions.   Another possibility would be
to give three different balls one unit of energy each $[(4 \times 3 \times 2)/(3 \times
2 \times 1) = 4$ complexions].   Finally, one ball could be given two units
and another ball one unit $(4 \times 3 = 12$ complexions).   The total complexions thus equal $4 + 4 + 12$, or 20 (see Table 4.2).

If one more unit of energy were added to the system there would be
five basic distributions.   All the energy could be given to one ball (4
complexions), each ball could be given one unit of energy (1 complexion),
two units could be given to each of two balls $(4 \times 3)/2 = 6$ complexions),
one ball could be given three units and another one unit $(4 \times 3 = 12$
complexions), or two balls could each be given one unit and another two
units $[(4 \times 3 \times 2)/2 = 12$ complexions].   Thus the addition of one
energy unit raises the total number of energy complexions from 20 to 35.
Randomness increases with the amount of available energy and hence
with *temperature*.

The overall randomness of the six straw–three energy unit system is
the product of the number of geometrical complexions (15) and the number of energy complexions (20).   This gives a total of 300 complexions.
That of the seven straw–four energy unit system is 35×35, or 1225
complexions.

The number of complexions is a property of the system just as is its
volume or energy.   One major difference exists, however.   Whereas
energy and volume are additive properties, the number of complexions is
multiplicative.   The total energy of a series of systems is the sum of the
energies of each of the individual systems, but the randomness is the
product of the number of complexions for each of the individual systems.
Randomness can be converted to an additive property in the following
way.   We will define a new property called *entropy*, designated S, which
is proportional to the logarithm of W:

$$S \propto \ln W \qquad \text{or} \qquad W \propto e^S$$

The number of complexions changes much more rapidly than the entropy.
Differentiating both sides of the expression, we obtain

$$dS \propto \frac{dW}{W}$$

TABLE 4.2

**Energy Complexions for a Single
Geometric Complexion†
(Total energy = 3 units)**

		1	2	3	4	5	6
	1	3	0	0	0	0	0
Mode	2	0	3	0	0	0	0
1	3	0	0	3	0	0	0
	4	0	0	0	3	0	0
	5	1	1	1	0	0	0
Mode	6	1	1	0	1	0	0
2	7	1	0	1	1	0	0
	8	0	1	1	1	0	0
	9	2	1	0	0	0	0
	10	2	0	1	0	0	0
	11	2	0	0	1	0	0
	12	1	2	0	0	0	0
	13	0	2	1	0	0	0
Mode	14	0	2	0	1	0	0
3	15	1	0	2	0	0	0
	16	0	1	2	0	0	0
	17	0	0	2	1	0	0
	18	1	0	0	2	0	0
	19	0	1	0	2	0	0
	20	0	0	1	2	0	0

† First four straws contain one ball; last two are empty: arrangement 1 in Table 4.1.

Entropy change is thus proportional to the fractional change in the number of complexions of the system.

Consider two systems, A and B. The total randomness is

$$W_{AB} = W_A W_B$$

The total entropy becomes

$$S_{AB} \propto \ln W_{AB}$$

$$\propto \ln (W_A W_B)$$

$$\propto \ln W_A + \ln W_B$$

$$= S_A + S_B$$

It proves convenient to assign to entropy the units of heat capacity; then TS has the units of energy. The gas constant, R, is selected as the proportionality constant.

$$dS = R \frac{dW}{W}$$

$$S = R \ln W$$

VARIATION OF ENTROPY WITH TEMPERATURE AT CONSTANT VOLUME

There are two ways in which the entropy of a substance can be raised: by increasing the energy associated with the motion of its constituent atoms or by increasing the free volume available to the atoms. Let us first consider how the number of complexions available to the atoms changes with temperature at constant volume.

Despite the fact that an infinite number of energy levels is available to each atom, the total energy restriction leads to a finite number of energy complexions. A major difference from the simple straw model given above must be taken into account, however, when dealing with atoms; the probability of occupation of the various energy levels is *not* the same for all levels. Boltzmann showed that the probability, p_i, of finding a given atom in any given energy level, ϵ_i, is proportional to exp $(-\epsilon_i/kT)$. The number of atoms at any given level drops exponentially with the energy of that level. Although a proof of Boltzmann's distribution law is beyond the scope of this book, a rather simple argument makes it plausible. The probability that an atom will gain energy during a collision is $\frac{1}{2}$. The mean energy change per mode of motion during a collision should be roughly equal to the mean energy per mode of motion, $kT/2$. Thus the probability that an atom will be raised to some unusually high energy, ϵ, is

$$p = \left(\tfrac{1}{2}\right)^{2\epsilon/kT}$$

Taking the natural logarithm of both sides,

$$\ln p = \frac{2\epsilon}{kT} \times \ln \frac{1}{2} \cong \frac{-\epsilon}{kT}$$

Taking the antilogarithm,

$$p \cong \exp \frac{-\epsilon}{kT}$$

As will be discussed in detail later in this chapter, there are only certain discrete energies which can be assumed by an atom for each of its modes of motion. For each direction of translational motion the energy levels are given by

$$\epsilon_n = n^2 \epsilon^0$$

where n is any integer and ϵ^0 is the lowest energy state. For rotational motion the levels are given by

$$\epsilon_n = n(n + 1)\epsilon^0$$

and for vibration by

$$\epsilon_n = (n + \tfrac{1}{2})\epsilon^0$$

For translation, the spacing between energy levels is extremely small (that is, $\epsilon^0 \ll kT$). In such a situation the Boltzmann occupation number is very nearly unity for all energy levels less than $\tfrac{1}{4}kT$ and very nearly zero for all energy levels above $4kT$. For energy levels in the range $\tfrac{1}{4}kT$ to $4kT$, the occupation number drops from unity to zero. This relationship is shown in Fig. 4.2. As can be seen in this figure, the situation may be crudely approximated by assuming that all energy levels with an energy less then kT have the same probability of being occupied and those with energies above kT have no probability of being occupied. For a single atom the number of energy complexions would then be equal to the number of these equally available levels. Hence for each mode of translational motion the number of energy complexions for a given atom is given by

$$W = \sqrt{\frac{kT}{\epsilon^0}}$$

The change in the number of complexions, dW, caused by a change in

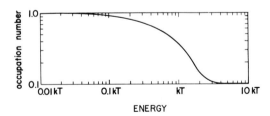

Fig. 4.2. Boltzmann occupation numbers for energy levels as a function of energy. . See text for detailed discussion.

temperature, dT, would be

$$dW = \sqrt{\frac{k}{\epsilon^0}} \frac{dT}{2\sqrt{T}}$$

and

$$\frac{dW}{W} = \frac{1}{2}\frac{dT}{T}$$

Since $dS = R\, dW/W$,

$$dS = \frac{R}{2}\frac{dT}{T}$$

From our considerations of how energy varies with temperature we know that $R/2$ is the heat capacity per mode of motion for 1 mole of atoms, and so

$$dS = C_V \frac{dT}{T}$$

The entropy change is equal to the energy added to the substance ($C_V\, dT$) divided by the temperature (T) at which it was added.   Hence energy added at 10°K is 10 times more effective in changing the entropy of translational motion than energy added at 100°K.

Where all three modes of translational motion are considered,

$$\frac{dW}{W} = \frac{3}{2}\frac{dT}{T}$$

and

$$dS = \frac{3R}{2}\frac{dT}{T}$$

As $3R/2$ is the translational heat capacity, again the entropy change is given by the heat added ($3R\, dT/2$) divided by the temperature at which it is added.

The same argument can be made for rotational motion.   The energy $\epsilon_n = n(n+1)\epsilon^0$ is the sum of the energies for the rotational modes in the nth rotational energy level.   Since n is very large, $n^2 \approx n(n+1)$ and $n \approx \sqrt{\epsilon_n/\epsilon^0}$.   For each mode of rotational motion the fractional change in energy complexions with temperature will be the same as for translational motion:

$$\frac{dW}{W} = \frac{1}{2}\frac{dT}{T}$$

and the entropy change per rotational mode for 1 mole of molecules
will be

$$dS = \frac{R}{2}\frac{dT}{T} = C_V \frac{dT}{T}$$

As will be shown below, this relationship is valid for all processes:

$$\left(\frac{dS}{dT}\right)_V = \frac{C_V}{T}$$

There is an important requirement, however, which we have not yet
stated.   Entropy changes can be directly related to heat addition only
if the process is reversible.   Since truly reversible processes are rare, it
would seem at first thought that this relationship of entropy to heat
capacity is not very useful.   The usefulness becomes apparent when we
remember that entropy is a *state function*, whose value depends only
on the state of the system and not on its history.   If we wish to calculate
the entropy change for a process, we simply specify the initial and final
states of the system and calculate the entropy change along a reversible
path connecting these states.

VARIATION OF ENTROPY WITH VOLUME AT CONSTANT TEMPERATURE

The number of complexions available to the atoms in any substance would
logically be related to the free volume (i.e., the volume not occupied by
the atoms themselves).   Hence, let us assume

$$W \propto V_{\text{free}} = fV$$

The change in the number of complexions must then be proportional to
the change in free volume,

$$dW \propto dV_{\text{free}} = dV$$

and thus the fractional change in complexions must equal the fractional
change in free volume,

$$\frac{dW}{W} = \frac{dV}{fV}$$

As shown in Chap. 2,

$$fV = \frac{RT}{P_{\text{thermal}}}$$

Substituting, we get

$$\frac{dW}{W} = \frac{P_{\text{thermal}}\, dV}{RT}$$

The entropy change per mole of atoms is then

$$dS = \frac{P_{\text{thermal}}\, dV}{T}$$

or the heat energy per mole ($P_{\text{thermal}}\, dV$) added to the substance divided by the temperature of addition.

Since $P_{\text{thermal}} = T(\alpha/\beta)$,

$$\left(\frac{\partial S}{\partial V}\right)_T = \frac{\alpha}{\beta}$$

The general relationship for entropy change in terms of temperature and volume changes becomes

$$dS = \frac{C_V}{T}\, dT + \frac{\alpha}{\beta}\, dV$$

By combining this relationship with that obtained for internal energy,

$$dE = C_V\, dT + \left(\frac{\alpha}{\beta}\, T - P\right) dV$$

the entropy change may be written in terms of internal energy and volume changes.   Thus

$$dS = \frac{1}{T}\, dE + \frac{P}{T}\, dV$$

ENTROPY CHANGE FOR VARIOUS TYPES OF PROCESSES

For an adiabatic reversible process the entropy change is zero, since an adiabatic change is, by definition, one in which no heat energy enters or leaves the system.

For an isothermal expansion the volume change is given by

$$dV = -\beta V\, dP$$

Hence

$$dS = \frac{\alpha}{\beta} dV = -\alpha V \, dP$$

or, in differential form,

$$\left(\frac{\partial S}{\partial P}\right)_T = -\alpha V$$

Finally, if a substance is heated under reversible conditions at constant pressure,

$$dV = \alpha V \, dT$$

and

$$dS = \frac{C_V}{T} dT + \frac{\alpha}{\beta} dV = \left(C_V + \frac{TV\alpha^2}{\beta}\right)\frac{dT}{T} = \frac{C_P}{T} dT$$

Again the entropy change is equal to the added heat divided by the temperature of addition. In differential form,

$$\left(\frac{\partial S}{\partial T}\right)_P = \frac{C_P}{T}$$

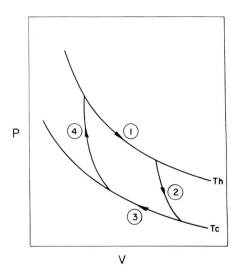

Fig. 4.3. Schematic representation of the Carnot cycle for an ideal gas.

THE CARNOT CYCLE

The famous ideal gas engine conceived by the nineteenth-century French physicist Carnot provides an example of the relationship between entropy change and heat energy added to the system. The reasoning runs as follows: For any cyclic process each time the system is returned to its initial state all its properties must also become equal to their initial values. In particular, the net entropy change for a complete cycle must be zero. Carnot considered a simple heat engine operated between two constant-temperature baths, one "hot," the other "cold." An ideal gas is carried around a four-stage cycle: (1) an isothermal expansion in contact with the hot reservoir, (2) an adiabatic expansion which cools the gas to the temperature of the cold reservoir, (3) an isothermal compression in contact with the cold reservoir, and (4) an adiabatic compression which heats the gas to its initial temperature. These steps are shown graphically in Fig. 4.3.

Since there is no entropy change associated with adiabatic processes, the entropy increase for the gas during the isothermal expansion must exactly equal the entropy decrease for the compression:

$$S_B - S_A = -(S_D - S_C)$$

The entropy change is given by

$$\Delta S = \frac{P \, \Delta V}{T} = \frac{R}{V} \Delta V$$

Thus

$$S_B - S_A = R \int_{V_A}^{V_B} \frac{dV}{V} = \ln V_B - \ln V_A$$

and

$$-(S_D - S_C) = -R \int_{V_C}^{V_D} \frac{dV}{V} = -\ln V_D + \ln V_C$$

This leads to the following relationship among the four volumes:

$$\ln V_B - \ln V_A = \ln V_C - \ln V_D$$

or

$$\frac{V_B}{V_A} = \frac{V_C}{V_D}$$

If the entropy principle is valid, this relationship must hold. We can see whether this is the case by considering the adiabatic changes. For such processes

$$\frac{C_V}{T} dT = -\frac{\alpha}{\beta} dV$$

For ideal gases

$$\frac{\alpha}{\beta} = \frac{P}{T} = \frac{R}{V}$$

Hence

$$\frac{C_V}{T} dT = -\frac{R}{V} dV$$

Thus for the adiabatic cooling from B to C,

$$\int_{T_H}^{T_C} \frac{C_V \, dT}{T} = -\int_{V_B}^{V_C} \frac{R}{V} dV$$

or

$$C_V \ln \frac{T_C}{T_H} = -R \ln \frac{V_C}{V_B}$$

and for the adiabatic heating from D to A,

$$C_V \ln \frac{T_H}{T_C} = -R \ln \frac{V_A}{V_D}$$

Hence

$$-R \ln \frac{V_C}{V_B} = R \ln \frac{V_A}{V_D}$$

and

$$\ln V_B - \ln V_C = \ln V_A - \ln V_D$$

or

$$\frac{V_B}{V_A} = \frac{V_C}{V_D}$$

Thus Carnot's cycle is consistent with the relationship among entropy, temperature, and volume given above.

ABSOLUTE ENTROPY

Since at absolute zero temperature almost all pure solid substances have only one geometric configuration ($W = 1$), the entropy of the substance is zero ($\ln 1 = 0$). Thus there can be no entropy change for any chemical reaction between these substances when carried out at absolute zero. For example, if calcite were converted to aragonite at absolute zero, the atoms would be shifted from one set of lattice points to another. Their randomness would not change. Since at absolute zero all vibrations are in their ground state there is also only one energy complexion. For this reason the absolute entropy of any pure substance can be defined as the entropy change it undergoes when taken from $T = 0°K$ (and $P = 1$) to that temperature and pressure of interest. The entropy of any substance can thus be computed from a knowledge of its heat capacity and coefficient of thermal expansion:

$$S = \int_0^T \frac{C_P}{T} dT - \int_1^P \alpha V \, dP$$

The variation of C_P and α with T and P must, of course, be taken into account in the integration.

The coefficient of thermal expansion, α, is zero at absolute zero, and so the choice of 1 atm for the heat-capacity integration is a matter of convenience chosen because C_P measurements are made at this pressure. This integration provides an accurate result only if the heat capacities are precisely known over the entire temperature range. Although small, the pressure effect on entropy becomes important beyond a few hundred atmospheres. The absolute entropy of any substance can in principle be obtained either by integrating measured heat capacities or by employing statistical mechanics.

For solids which undergo no phase transformations between absolute zero and the temperature of interest, the integration is relatively simple:

$$S = \int_{0_1 \text{ atm}}^T \frac{C_P}{T} dT - \int_{1_T}^P \alpha V \, dP$$

If phase changes take place, the entropy changes associated with them must be taken into account. For example, if a substance undergoes a solid-solid, then a solid-liquid, and finally a liquid-to-gas transformation, its entropy would be given by

$$S = \int_0^{T_{\text{sol}_1-\text{sol}_2}} \frac{(C_P)_{\text{sol}_1}}{T} dT + \frac{\Delta H_{\text{sol}_1-\text{sol}_2}}{T_{\text{sol}_1-\text{sol}_2}} + \int_{T_{\text{sol}_1-\text{sol}_2}}^{T_{\text{sol}_2-\text{liq}}} \frac{(C_P)_{\text{sol}_2}}{T} dT$$

$$+ \frac{\Delta H_{\text{sol}_2-\text{liq}}}{T_{\text{sol}_2-\text{liq}}} + \int_{T_{\text{sol}_2-\text{liq}}}^{T_{\text{liq}-\text{gas}}} \frac{(C_P)_{\text{liq}}}{T} dT + \frac{\Delta H_{\text{liq}-\text{gas}}}{T_{\text{liq}-\text{gas}}}$$

$$+ \int_{T_{\text{liq}-\text{gas}}}^T \frac{(C_P)_{\text{gas}}}{T} dT - \int_1^P \alpha V \, dP$$

For an ideal Debye solid the entropy will show a simple relationship to the characteristic temperature. The higher the value of θ_D, the lower the entropy at any given temperature. This relationship is shown in Fig. 4.4. Table 4.3 shows that solids with low heat capacities also have low entropies.

TABLE 4.3

Entropy and Heat Capacity per Atom for Several Solid Compounds†

Compound	$S,$ cal/deg-mole	S/n	$C_P/n,$ at 298°K
BeO	3.37	1.68	3.03
SiO$_2$ (α quartz)	10.00	3.33	3.54
Al$_2$SiO$_5$ (kyanite)	20.02	2.50	3.64
CaCO$_3$ (calcite)	22.2	4.44	3.91
PbCO$_3$	31.3	6.3	4.18
TiO$_2$ (rutile)	13.16	4.01	4.39
Fe$_3$O$_4$	35.0	5.0	4.89
CaF$_2$	16.46	5.45	5.34

† Entropy data taken from K. K. Kelley and E. G. King, *U.S. Bur. Mines Bull.*, 592, 1961. Heat-capacity data taken from K. K. Kelley, *U.S. Bur. Mines Bull.*, 584, 1960.

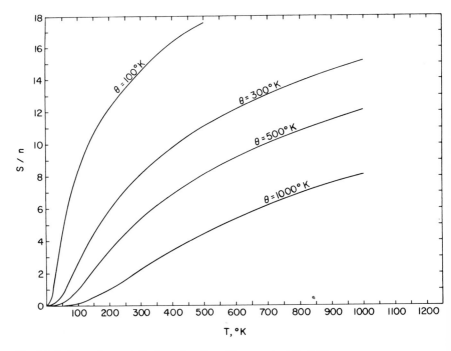

Fig. 4.4. Debye entropy plotted as a function of temperature for four values of θ_D. To obtain the molar entropy, multiply by the number of atoms in the compound.

To illustrate the effects of temperature and pressure on the entropy of a solid, we have calculated an entropy cycle for calcite and aragonite starting at room temperature and pressure. The results are shown in Fig. 4.5. Increasing the temperature from 25°C (298°K) to 225°C (498°K) results in a 50-percent increase in the entropies of calcite and aragonite. Despite these large changes, the difference in entropy between calcite and aragonite remains small. Raising the pressure from 1 to 5000 results in extremely small changes in the entropies of calcite and aragonite. For many calculations involving solids the effect of pressure on entropy may be neglected.

STATISTICAL MECHANICS

As briefly outlined above, from the Boltzmann distribution law and the restriction of the energy of atoms to certain discrete values it can be shown that the fractional increase in the number of energy complexions for any given mode of translational or rotational motion is equal to one-

CALCITE

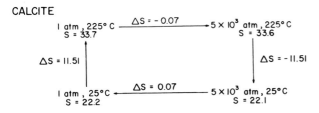

ARAGONITE

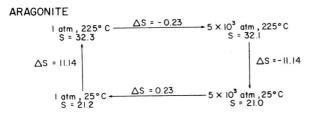

CALCITE - ARAGONITE

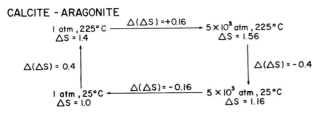

Fig. 4.5. Entropy cycle for calcite and aragonite. Values of thermodynamic properties used in the calculation are given in Table 3.6.

half the fractional increase in temperature. If entropies are to be obtained by statistical calculations, it is necessary to obtain a rigorous relationship between the number of energy complexions and temperature for all types of motion, including partially activated modes.

First let us see why the motions of atoms are restricted to discrete energy levels. This restriction is a reflection and consequence of the dual nature of matter. At the beginning of this century a group of scientists realized that many of the properties of rapidly moving particles could be understood only if the particles were considered to act as waves. For example, electrons are diffracted on passing through a crystal. The mechanism of diffraction is well understood for waves such as x-rays; the problem of electron diffraction is easily understood if we can endow electrons with the properties of waves.

In 1923, the physicist Duc Louis de Broglie proposed that all matter exhibited this dual nature. We do not observe diffraction or other wave effects in the case of particles with masses of a gram or so because they have such small wavelengths. However, the wave nature of the particle is still present. The wavelength, λ, of a "particle wave" is called the *de Broglie wavelength* of the particle. It is equal to Planck's constant, h, divided by the momentum of the particle, mv:

$$\lambda = \frac{h}{mv}$$

We may easily obtain the de Broglie wavelength by starting with Einstein's principle of mass-energy conversion. If a particle of mass, m, were converted entirely into a quantum (or "bundle") of electromagnetic radiation, the wavelength of this quantum would be found as follows: Einstein's equation states that the energy release, E, is equal to the mass of the particle times the speed of light squared ($E = mc^2$). This energy must equal the energy of the radiation which is given by $h\nu$ or hc/λ. Thus,

$$mc^2 = \frac{hc}{\lambda} \qquad \text{or} \qquad \lambda = \frac{h}{mc}$$

The quantity mc represents the momentum of the photon, and so

$$\lambda = \frac{h}{p}$$

De Broglie maintained that this equation is true for all particles, if p is the momentum of the particle.

If a moving particle exhibits wave properties, it must always move so that the waves it produces are in phase. This produces constructive interference. Destructive interference, or out-of-phase waves, is not allowed. For a rotating molecule, constructive interference can occur only when its rotation frequency is an exact multiple of the fundamental frequency of its de Broglie wave. This condition obtains only at certain critical values of the rotation energy. These values represent the allowed energy levels. Other rotation energies are excluded because of destructive interference.

In 1926, Erwin Schrödinger formulated a differential equation from which quantum energy levels could be calculated. He started with the differential equation which defines all wave motion, which in one dimension is

$$\frac{\partial^2 A}{\partial x^2} = \frac{1}{v^2}\frac{\partial^2 A}{\partial t^2}$$

To separate out the time dependence, he defined a parameter ψ as a periodic function of time:

$$\psi = \frac{A}{\sin 2\pi(v/\lambda)t}$$

In terms of ψ the wave equation becomes

$$\frac{d^2\psi}{dx^2} + \frac{4\pi^2}{\lambda^2}\psi = 0$$

Substituting the de Broglie value for λ, we obtain

$$\frac{d^2\psi}{dx^2} + \frac{4\pi^2 m^2 v^2}{h^2}\psi = 0$$

Since the kinetic energy of the particle, $mv^2/2$, is equal to its total energy, ϵ, minus its potential energy, u, we have

$$\frac{d^2\psi}{dx^2} + \frac{8\pi^2 m}{h^2}(\epsilon - u)\psi = 0$$

This is the one-dimensional form of Schrödinger's famous equation. This equation is as important to quantum mechanics as Newton's $F = ma$ is to classical mechanics. We shall now consider the results obtained from the solutions to this equation for various types of motion.

DETERMINATION OF ENERGY LEVELS

Potential energy has the agreeable property that the zero point may be arbitrarily fixed. It is convenient to choose the translational potential energy of a molecule moving in a gas to be zero. The Schrödinger equation for one dimension then becomes

$$\frac{d^2\psi}{dx^2} + \frac{8\pi^2 m\epsilon}{h^2}\,\psi = 0$$

In an ideal gas, the molecules can be considered to collide with only the walls of the container and not with each other. The allowed values of the particle's energy may then be calculated from the Schrödinger equation. The result is

$$\epsilon_n = \frac{n^2 h^2}{8ma^2}$$

where n may be any positive integer, m is the mass of the particle, and a is the distance between the walls of the container. The spacing between energy levels decreases with increasing mass of the particle and with increasing size of the container. The energy spacing between adjacent levels increases with the number of the level:

$$\Delta\epsilon_{n+1,n} = (2n + 1)\,\frac{h^2}{8ma^2}$$

The spacing between level 4 and level 5 is 3 times the spacing between levels 1 and 2.

If the problem is solved for an ideal gas in a real three-dimensional container, the result has three quantum numbers, n_1, n_2, and n_3. These quantum numbers are associated with the three orthogonal dimensions of the container, a_1, a_2, and a_3:

$$\epsilon_n = \frac{h^2}{8m}\left(\frac{n_1^2}{a_1^2} + \frac{n_2^2}{a_2^2} + \frac{n_3^2}{a_3^2}\right)$$

This solution allows two different states to have the same energy, a condition called *degeneracy*. For example, if the container is a cube with $a_1 = a_2 = a_3$, a threefold degeneracy is possible for the energy $\epsilon = 6h^2/8ma^2$. This energy results from

(1)	$n_1 = 1$	$n_2 = 1$	$n_3 = 2$
(2)	$n_1 = 1$	$n_2 = 2$	$n_3 = 1$
(3)	$n_1 = 2$	$n_2 = 1$	$n_3 = 1$

To calculate the rotational energy levels, we may consider a diatomic molecule to be a dumbbell with masses at each end of a rod (the molecular bond axis). The potential energy of this rotating dumbbell is taken as zero. We may further simplify the calculations by reducing the problem to one in which a single object of mass, μ, rotates at a distance, r, around an origin. This distance is the distance of separation of the original masses. If the angular momentum of the two systems is to be the same, the masses must satisfy the condition

$$\mu = \frac{m_1 m_2}{m_1 + m_2}$$

The quantity μ is called the *reduced mass* of the system. For this case the Schrödinger equation becomes

$$\frac{d^2\psi}{dx^2} + \frac{8\pi^2\mu\epsilon}{h^2}\psi = 0$$

$$\epsilon_j = \frac{j(j+1)h^2}{8\pi^2\mu r^2} = \frac{j(j+1)h^2}{8\pi^2 r^2[m_1 m_2/(m_1 + m_2)]}$$

where j is any positive integer (or zero), m_1 and m_2 are the masses of the atoms in the diatomic molecule, and r is the bond length. The separation of energy levels again decreases with increasing mass of either atom making up the molecule. It also decreases with increasing bond length. The spacing between levels increases with increasing level number:

$$\Delta\epsilon_{n+1,n} = 2(n+1)\frac{h^2}{8\pi^2\mu r^2}$$

We may calculate the vibrational energy levels of a diatomic molecule by using a simple idealized model. This model will consist of two masses connected by a spring so that the "molecule" resembles a dumbbell. We shall assume that the restoring force exerted on the masses is proportional to the linear distortion of the spring. This distortion is the deviation of the separation between the end masses from the equilibrium separation distances. Our assumption could be more concisely stated by claiming that the spring obeys Hooke's law.

The motion of an oscillator consisting of two masses joined by a spring is analogous to that of a ball rolling in a parabolic track, as shown in Fig. 4.6. The potential energy of the ball at point (1) is given by mgh. For a parabolic track, the elevation, h, is proportional to the square of the horizontal distance, x, from the null point. This means that at the top of the track the energy of the ball is proportional to x^2. The energy of the vibrating dumbbell at one of the extremes of its motion is $\frac{1}{2}kx^2$. The constant k is a measure of the rigidity of the spring (or molecular bond) and x is the difference between the separation of the ends at equilibrium and at furthest extension or compression. At the bottom of the parabolic track, the velocity of the ball is a maximum. Similarly, when the vibrating system passes through its equilibrium separation distance, the velocity of the end masses is a maximum.

We may, as for the case of rotation, further simplify the model and consider a single mass at the end of a rigidly mounted spring. This single mass would be the reduced mass of the isolated system, $m_1 m_2/(m_1 + m_2)$. The Schrödinger equation for this case is

$$\frac{d^2\psi}{dx^2} + \frac{8\pi^2[m_1 m_2/(m_1 + m_2)]}{h^2}\left(\epsilon - \frac{1}{2}kx^2\right)\psi = 0$$

Real solutions are obtained only if

$$\epsilon = \left(n + \frac{1}{2}\right)\frac{h}{2\pi}\sqrt{\frac{k}{m_1 m_2/(m_1 + m_2)}}$$

where n is any positive integer or zero. The spacing between levels again decreases with an increase in the mass of either atom. The spacing also decreases with decreasing rigidity of the bond. The energy levels in the harmonic approximation which we have used are equally spaced:

$$\Delta\epsilon = \frac{h}{2\pi}\sqrt{\frac{k}{m_1 m_2/(m_1 + m_2)}}$$

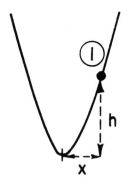

Fig. 4.6. Ball rolling in a parabolic track, and its associated potential energy.

As mentioned in Chap. 3, the spacing between energy levels is greatest for vibration and least for translation.   Let us consider as an example the molecule CO, carbon monoxide.   The force constant, k, for CO is 1.85×10^6 dynes/cm; the reduced mass, μ, is 1.14×10^{-23} g; and $h/2\pi$ is 1.054×10^{-27} erg-sec.   The spacing of vibrational levels is

$$\Delta\epsilon_{\text{vib}} = (1.054 \times 10^{-27}) \sqrt{\frac{1.85 \times 10^6}{1.14 \times 10^{-23}}}$$

$$\Delta\epsilon_{\text{vib}} = 4.25 \times 10^{-13} \text{ erg}$$

This energy difference is large when compared with the average thermal energy of a particle, kT, at room temperature.   For 300°K,

$$kT = (1.38 \times 10^{-16}) \times (300)$$

$$= 4.14 \times 10^{-14} \text{ erg}$$

The spacing between rotational levels for CO can be calculated using $r = 1.13 \times 10^{-8}$ cm.   This value is obtained from analysis of infrared rotational spectra.   The spacing between the levels $j = 1$ and $j = 0$ is

$$\Delta\epsilon_{1,0} = \frac{2h^2}{8\pi^2 \mu r^2}$$

$$= 1.53 \times 10^{-15} \text{ erg}$$

This spacing is very small compared with kT at room temperature.   The spacing $\Delta\epsilon_{1,0}$ would be equal to kT at 11.1°K; this temperature is well below the solidification temperature of CO.

Let us calculate the spacing between translational levels for a CO molecule in a cubic container 30 cm on a side.   The volume of this container would be 27 liters.   The spacing between levels 1 and 2 is given by

$$\Delta\epsilon_{2,1} = \frac{(2 + 1)(6.62 \times 10^{-27})^2}{(8)(4.65 \times 10^{-23})(9 \times 10^2)}$$

$$= 3.93 \times 10^{-34} \text{ erg}$$

This value is infinitesimally small compared with kT at any reasonable temperature.

The form of the energy-level quantizations for the three modes of motion can be seen without becoming involved with the complex Schrödinger equation.   For the case of translation we may consider a ball bouncing between the ends of a closed cylindrical tube. If the wave associated with this ball is to exist, the half wavelength, $\lambda/2$, must be some integral factor of the length, l, of the tube:

$$\frac{\lambda}{2} = \frac{l}{n}$$

Applying de Broglie's equation,

$$p = \frac{h}{\lambda} = \frac{nh}{2l}$$

Since

$$\epsilon = \frac{mv^2}{2} = \frac{p^2}{2m}$$

we find that energy is quantized and

$$\epsilon = \frac{n^2h^2}{8ml^2}$$

This is exactly the result obtained by solving the Schrödinger equation for the one-dimensional case.

To examine the case of rotation, let us imagine a ball rotating in a circular orbit at the end of a string. The associated de Broglie wavelength, λ, must be some integral factor of the circumference of the circular orbit:

$$\lambda = \frac{2\pi r}{n}$$

From the de Broglie relation,

$$p = \frac{h}{\lambda} = \frac{nh}{2\pi r}$$

The mass at the end of the string represents the reduced mass, μ, of a diatomic rotating molecule, and so

$$\epsilon = \frac{\mu v^2}{2} = \frac{p^2}{2\mu}$$

$$= \frac{n^2h^2}{8\pi^2\mu r^2}$$

Again we find that the energy levels are quantized, but we have a slightly different result from that obtained by using the Schrödinger equation. The discrepancy is caused by the fact that atoms are not "classical particles."

The vibrational quantization is obtained by requiring that all vibrational modes must occur at natural resonant frequencies. Thus

$$\epsilon = nh\nu = \frac{nh}{2\pi}\sqrt{\frac{k}{\mu}}$$

We again have a result which is close to, but slightly different from, the Schrödinger value. Our problem now is that we allow the vibrator to be at rest; quantum mechanics denies this. Even at absolute zero temperature the vibrator must possess energy. This energy,

$$\epsilon_0 = \tfrac{1}{2}h\nu$$

is called the *zero-point energy* of the system.

PARTITION FUNCTIONS

Now that the energy levels available to an atom are known, we must inquire into how the atoms will distribute themselves among the available levels. For any one atom, the Boltzmann distribution gives the fraction of time, f_i, the atom spends in any particular level, i,

$$f_i = \frac{\exp\left(-\epsilon_i/kT\right)}{\sum_j \exp\left(-\epsilon_j/kT\right)}$$

The summation appearing in the denominator is given the symbol q and is defined as the single-particle partition function.

We may now calculate the partition functions for the various types of motion. First we will consider translation. For each direction of motion, each molecule will have an associated partition function given by

$$q_{Tx} = \sum_n \exp\left(-\frac{n^2h^2}{8ma^2kT}\right)$$

Except at temperatures extremely close to absolute zero, this summation may be evaluated as an integral, yielding

$$q_{Tx} = \sqrt{\frac{kT}{h^2/2\pi m a^2}}$$

The lowest energy level is

$$\epsilon_1 = \frac{h^2}{8ma^2}$$

and so we may write the translational partition function as follows:

$$q_{Tx} = \frac{\sqrt{\pi}}{2}\sqrt{\frac{kT}{\epsilon_1}}$$

Since q_{Tx} is a measure of the average number of energy complexions available to each mode of translational motion, this result confirms the validity of the approximation made earlier in the chapter where we claimed that the number of complexions was approximately n. The total partition function for translation is

$$q_{\text{trans}} = q_{Tx}q_{Ty}q_{Tz} = \left(\frac{2\pi mkT}{h^2}\right)^{\frac{3}{2}} V$$

where $V = a^3$.

The rotational partition function is slightly more complicated since a rotating diatomic molecule has two axes of rotation. The available rotational energy can be distributed in $2j + 1$ ways between these two axes. If, for example, $j = 1$, there are $j(j + 1)$, or 2, quanta of energy available to the system. One can be given to each axis, two can be given to the first axis and none to the second, or two to the second and none to the first. This is a total of $2j + 1$, or 3, combinations. Since there are $2j + 1$ states of the system with energy ϵ_j, the levels are said to have $(2j + 1)$-fold degeneracy.

The partition function, which is a sum over *states* of the system, is

$$q_{\text{rot}} = \sum_j (2j + 1) \exp\left[-\frac{j(j + 1)h^2}{8\pi^2\mu r^2 kT}\right]$$

For temperatures at which kT exceeds the spacing between levels, this summation yields†

$$q_{rot} = \frac{kT}{h^2/8\pi^2\mu r^2} = \frac{8\pi^2\mu r^2 kT}{h^2}$$

Since the $j = 1$ energy level, ϵ_1, has energy

$$\epsilon_1 = \frac{2h^2}{8\pi^2\mu r^2}$$

we may express the total rotational partition function in terms of the energy of this level:

$$q_{rot} = \frac{2kT}{\epsilon_1}$$

We may approximate the contribution of each rotational mode separately, q'_{rot}, by taking the square root of q_{rot}:

$$q'_{rot} = \sqrt{\frac{2kT}{\epsilon_1}}$$

The partition function for vibration must be treated in a manner different from that for translation and rotation. This is because the temperature at which the spacing between levels equals kT is generally several hundred degrees. For CO, this temperature would be 3080°K. Thus the assumption that many levels are populated at room temperature, which is valid for translation and rotation, is clearly invalid for vibration. Almost all the CO molecules remain in their ground vibration state at room temperature.

The partition function for vibration is given by

$$q_{vib} = \sum_n \exp\left[-\frac{(n + \frac{1}{2})h\nu}{kT}\right]$$

Since the levels are widely spaced, we may not integrate; fortunately this

† There is one more detail which must be considered to make this equation completely accurate. For homonuclear molecules like O_2 or N_2, the integers n must be either all even or all odd. This is referred to as a symmetry restriction. Half the states available to heteronuclear molecules like CO or NO are forbidden to homonuclear molecules. To allow for this restriction, q_{rot} for homonuclear molecules must contain an additional factor of 2 in the denominator.

expression may be written

$$q_{\text{vib}} = \frac{\exp\left(-h\nu/2kT\right)}{1 - \exp\left(-h\nu/kT\right)}$$

At room temperature

$$\exp\left(-\frac{h\nu}{kT}\right) \ll 1$$

and

$$q_{\text{vib}} \cong \exp\left(-\frac{h\nu}{2kT}\right)$$

To underscore the vast difference between population of vibrational levels and population of rotational levels, let us calculate the distribution of atoms among the first three vibrational levels for CO. The partition function is

$$q_{\text{vib}} = \frac{\exp\left(-h\nu/2kT\right)}{1 - \exp\left(-h\nu/kT\right)}$$

For CO,

$$h\nu = 4.26 \times 10^{-13} \text{ erg}$$

$$kT = 4.13 \times 10^{-14} \text{ erg}$$

at 300°K. Thus

$$q_{\text{vib}} = \frac{0.0058}{1 - 0.000033}$$

The Boltzmann distribution gives the fraction of atoms in each level, i:

$$f_i = \frac{\exp\left(-\epsilon_i/kT\right)}{q_{\text{vib}}}$$

and the energy of each level is given by

$$\epsilon_i = (i + \tfrac{1}{2})h\nu$$

For the ground state, $i = 0$, and

$$\epsilon_0 = \tfrac{1}{2}h\nu$$

Thus the fraction of atoms in the ground state is

$$f_0 = \frac{\exp(-h\nu/2kT)}{q_{\text{vib}}}$$

$$= 1 - 0.000033$$

$$\cong 0.99997$$

Only about three atoms in every one hundred thousand are not in the ground state.

For the first excited state, ϵ_1,

$$\epsilon_1 = \tfrac{3}{2}h\nu$$

and

$$f_1 = \frac{\exp(-3h\nu/2kT)}{q_{\text{vib}}}$$

$$\cong \frac{1.9 \times 10^{-7}}{5.8 \times 10^{-3}} = 3.28 \times 10^{-5}$$

Essentially all the atoms which are not in the ground state are in the first excited state.   The second excited state with energy

$$\epsilon_2 = \tfrac{5}{2}h\nu$$

contains only about one atom in 10^9.

The total partition function for a single diatomic molecule in a box of volume V is

$$q = q_{\text{trans}} q_{\text{rot}} q_{\text{vib}}$$

where

$$q_{\text{trans}} = \left(\frac{2\pi mkT}{h^2}\right)^{\frac{3}{2}} V$$

$$q_{\text{rot}} = \frac{8\pi^2 \mu r^2 kT}{h^2}$$

$$q_{\text{vib}} = \frac{\exp(-h\nu/2kT)}{1 - \exp(-h\nu/kT)}$$

Since the translational term incorporates volume changes, this q is a measure of the total (volume + energy) complexions available to the molecule.

If more than one molecule is present in the box, then the q for each molecule becomes

$$q = \left[\frac{(q_{\text{trans}}q_{\text{rot}}q_{\text{vib}})^N}{N!} \right]^{1/N}$$

or,

$$q = \frac{q_{\text{trans}}q_{\text{rot}}q_{\text{vib}}}{(N!)^{1/N}}$$

The $N!$ factor must be introduced for the same reason as it was in the straw problem.   The molecules are indistinguishable.

For large numbers,

$$(N!)^{1/N} = \frac{e}{N}$$

Hence

$$q = \frac{q_{\text{trans}}q_{\text{rot}}q_{\text{vib}}e}{N}$$

For purposes of calculation, it is convenient to combine the term resulting from indistinguishability, e/N, with the translational partition function:

$$q_{TI} = \left(\frac{2\pi mkT}{h^2} \right)^{\frac{3}{2}} \frac{V e}{N}$$

From a knowledge of the partition function of a gas molecule it is possible to calculate the absolute entropy of the gas.   As shown above, at constant volume

$$dS = \frac{C_V}{T} dT$$

Hence

$$S = \int_0^T \frac{C_V}{T} dT$$

By definition

$$C_V = \left(\frac{\partial E}{\partial T} \right)_V$$

The internal energy can be calculated from the partition function as follows: The total energy of the system is the sum of the contribution due to each molecule:

$$E = N\Sigma f_i \epsilon_i$$

where ϵ_i represents various total energies that a molecule can assume.

In accordance with the discussion above,

$$f_i = \frac{\exp\left(-\epsilon_i/kT\right)}{\Sigma \exp\left(-\epsilon_i/kT\right)} = \frac{\exp\left(-\epsilon_i/kT\right)}{q}$$

Hence

$$E = \frac{N}{q} \sum \epsilon_i \exp\left(-\frac{\epsilon_i}{kT}\right)$$

However,

$$\sum \epsilon_i \exp\left(-\frac{\epsilon_i}{kT}\right) = kT^2 \frac{dq}{dT}$$

Thus

$$E = RT^2 \frac{1}{q}\frac{dq}{dT}$$

$$C_V = \frac{\partial}{\partial T} RT^2 \frac{1}{q}\frac{dq}{dT}$$

and

$$S = \int_0^T \frac{1}{T}\frac{\partial}{\partial T} RT^2 \frac{1}{q}\frac{dq}{dT}\, dT$$

Integrating by parts,

$$S = RT \frac{1}{q}\frac{dq}{dT} + R \int_0^T \frac{1}{q}\frac{dq}{dT}\, dT$$

Writing the remaining integral in terms of q rather than T, since $q \to 1$ as $T \to 0$,

$$S = RT \frac{1}{q}\frac{dq}{dT} + R \int_1^q \frac{1}{q}\, dq$$

which yields

$$S = RT \frac{1}{q}\frac{dq}{dT} + R \ln q$$

or

$$S = RT \frac{d\ln q}{dT} + R \ln q$$

We have shown that for a gas,

$$q = q_{TI} q_{\text{rot}} q_{\text{vib}}$$

Hence

$$\ln q = 1 + \ln \frac{q_{\text{trans}}}{N} + \ln q_{\text{rot}} + \ln q_{\text{vib}}$$

Thus

$$S = R + RT \frac{d \ln (q_{\text{trans}}/N)}{dT} + R \ln \frac{q_{\text{trans}}}{N}$$

$$+ RT \frac{d \ln q_{\text{rot}}}{dT} + R \ln q_{\text{rot}}$$

$$+ RT \frac{d \ln q_{\text{vib}}}{dT} + R \ln q_{\text{vib}}$$

TRANSLATIONAL AND INDISTINGUISHABILITY ENTROPY

The entropy of a monatomic gas would have only the component due to translation and indistinguishability:

$$S_{TI} = R + \frac{RT \, d \ln (q_{\text{trans}}/N)}{dT} + R \ln \frac{q_{\text{trans}}}{N}$$

As stated above,

$$\frac{q_{\text{trans}}}{N} = \left(\frac{2\pi m k}{h^2}\right)^{\frac{3}{2}} \frac{V}{N} T^{\frac{3}{2}}$$

$$\frac{d(q_{\text{trans}}/N)}{dT} = \left(\frac{2\pi m k}{h^2}\right)^{\frac{3}{2}} \frac{V}{N} \frac{3}{2} \sqrt{T}$$

Hence

$$RT \frac{d \ln (q_{\text{trans}}/N)}{dT} = \frac{3}{2} R$$

This yields

$$S_{TI} = \tfrac{5}{2} R + R \ln \left(\frac{2\pi m k T}{h^2}\right)^{\frac{3}{2}} \frac{V}{N}$$

The entropy of a monatomic gas then rises by an amount $\frac{3}{2}R$ for each factor of e ($\sim$2.7) the temperature is increased and by an amount R for each 2.7-fold increase in molar volume. If the atoms of gas A were 2.7 times as heavy as those of gas B (Ne and He differ by about this factor) gas A would have an entropy $\frac{3}{2}R$ greater than gas B.

The translation entropy of 1 mole of CO at 300°K in a volume of

24.6 liters (2.46×10^4 cm³) can be calculated as follows: Under these conditions the pressure of the gas is 1 atm.   The values of the parameters required are

$$m = \frac{28 \text{ g}}{\text{mole}} \frac{1 \text{ mole}}{6.02 \times 10^{23} \text{ molecules}}$$

$$N = 6.02 \times 10^{23} \text{ molecules}$$

$$kT = 4.14 \times 10^{-14} \text{ erg}$$

$$h^2 = 4.38 \times 10^{-53} \text{ erg}^2\text{-sec}^2$$

$$S_{TI} = \tfrac{5}{2}R + R \ln \left(\frac{2\pi m k T}{h^2}\right)^{\tfrac{3}{2}} \frac{V}{N}$$

$$= \tfrac{5}{2}R + R \ln \left\{ \left[\frac{(6.28)(2.8 \times 10^1)(4.14 \times 10^{-14})}{(4.38 \times 10^{-53})(6.02 \times 10^{23})} \right]^{\tfrac{3}{2}} \frac{2.46 \times 10^4}{6.02 \times 10^{23}} \right\}$$

$$= \tfrac{5}{2}R + R \ln (5.92 \times 10^6)$$

$$= \tfrac{5}{2}R + 2.303R \log (5.92 \times 10^6)$$

$$= 18.1R$$

$$= 36.0 \text{ cal/deg-mole}$$

ROTATIONAL ENTROPY

The rotational contribution to the entropy of a diatomic gas is

$$S_{\text{rot}} = RT \frac{d \ln q_{\text{rot}}}{dT} + R \ln q_{\text{rot}}$$

As shown above,

$$q_{\text{rot}} = \frac{8\pi^2 \mu r^2 k T}{h^2}$$

and thus

$$\frac{d q_{\text{rot}}}{dT} = \frac{8\pi^2 \mu r^2 k}{h^2}$$

and

$$RT \frac{d \ln q_{\text{rot}}}{dT} = \frac{RT}{q_{\text{rot}}} \frac{d q_{\text{rot}}}{dT} = R$$

Hence

$$S_{\text{rot}} = R + R \ln \frac{8\pi^2 \mu r^2 k T}{h^2}$$

For CO at 1 atm, 300°,

$$\mu = 1.14 \times 10^{-23} \text{ g}$$
$$r = 1.13 \times 10^{-8} \text{ cm}$$
$$\mu r^2 = 1.45 \times 10^{-39} \text{ g-cm}^2$$
$$S_{\text{rot}} = R + R \ln 108 = 5.7R$$
$$= 11.3 \text{ cal/deg-mole}$$

Since, as we shall see below, the vibrational contribution to the entropy of CO is negligible at room temperature,

$$S_{\text{CO}}^{\text{std}} = 36.0 + 11.3 = 47.3 \text{ cal/deg-mole}$$

VIBRATIONAL CONTRIBUTION TO ENTROPY

As shown above, the vibrational entropy is given by

$$S_{\text{vib}} = RT \frac{d \ln q_{\text{vib}}}{dT} + R \ln q_{\text{vib}}$$

and

$$q_{\text{vib}} = \frac{\exp\left(-h\nu/2kT\right)}{1 - \exp\left(-h\nu/kT\right)}$$

Taking the derivative of q_{vib} with respect to T,

$$\frac{dq_{\text{vib}}}{dT} = \frac{(h\nu/2kT^2) \exp\left(-h\nu/2kT\right)}{1 - \exp\left(-h\nu/kT\right)}$$
$$+ \frac{\exp\left(-h\nu/2kT\right)[\exp\left(-h\nu/kT\right)](h\nu/kT^2)}{[1 - \exp\left(-h\nu/kT\right)]^2}$$

$$\frac{1}{q_{\text{vib}}} \frac{dq_{\text{vib}}}{dT} = \frac{h\nu}{2kT^2} + \frac{[\exp\left(-h\nu/kT\right)](h\nu/kT^2)}{1 - \exp\left(-h\nu/kT\right)}$$

Thus the vibrational entropy is

$$S_{\text{vib}} = \frac{Rh\nu}{2kT} + \frac{Rh\nu}{kT} \frac{\exp\left(-h\nu/kT\right)}{1 - \exp\left(-h\nu/kT\right)} + R \ln \frac{\exp\left(-h\nu/2kT\right)}{1 - \exp\left(-h\nu/kT\right)}$$

$$= \frac{Rh\nu}{kT} \frac{1}{\exp\left(h\nu/kT\right) - 1} - R \ln\left[1 - \exp\left(-\frac{h\nu}{kT}\right)\right]$$

For CO at 300°K,

$$\frac{h\nu}{kT} = 10.3$$

$$\exp - \frac{h\nu}{kT} = 3.3 \times 10^{-5}$$

$$S_{\text{vib}} = -R \ln [1 - (3.3 \times 10^{-5})] + R(10.3) \frac{1}{e^{10.3} - 1}$$

The first term is very close to zero. The value of the second term is

$$S_{\text{vib}} = 7.3 \times 10^{-4} \text{ cal/deg-mole}$$

The vibrational term contributes a negligible amount at low temperatures but becomes very important at high temperatures.

This concludes our discussion of statistical mechanics. The results we have derived are very basic and show only a small part of the power of the method. The reader who is interested in learning more about this subject is referred to an excellent book by Frank C. Andrews entitled "Equilibrium Statistical Mechanics" (John Wiley & Sons, Inc., New York, 1963).

PROBLEMS

4.1 A box has five equally spaced platforms arranged in a vertical series. Let the first platform represent one energy unit, the second, two units, etc. Six marbles, three of which are white and three of which are black, are placed in the box. What is the total number of complexions for a total energy of 12 units?

4.2 Find the de Broglie wavelength for (a) a golf ball weighing 50 g and traveling at 100 m/sec; (b) a $\frac{1}{25}$-Mev neutron (thermal neutron at $T \approx 300°K$). Would you expect to see any wave effects from either particle?

4.3 A box has sides a_1, a_2, a_3 and contains one particle. If $a_2 = a_3 = 2a_1$, find the first four degenerate energy levels for translation. What is the order of the degeneracy?

4.4 The entropy of NaCl (solid) at 25°C is 17.30 cal/deg-mole; its heat capacity is $C_P = 10.98 + (3.90 \times 10^{-3} T, °K)$ cal/deg-mole. The density at 1 atm, 25°C is 2.163 g/cm³. The average coefficient of thermal expansion from 20 to 200°C is 1.26×10^{-4} deg^{-1}; the compressibility is 4.26×10^{-6} atm^{-1}. Calculate the entropy at 100 atm, 200$_{o}$C.

4.5 The bond length for HCl^{35} is 1.30 Å; its fundamental wave number for vibration is 2989 cm^{-1}. Calculate the statistical entropy of HCl^{35} at 300°K, 1 atm.

4.6 Show that if q_{Tx} gives the average number of translational states for the x component of direction, the number of these states is approximately equal to n, the quantum number of the state whose energy equals the mean thermal energy for that component.

chapter five Equilibrium and Its Attainment

In any chemical problem two relevant questions arise. The first is, What is the equilibrium configuration of the atoms present? This question can be answered by thermodynamics. The second is, How long will it take for this configuration to be established? Here, thermodynamics cannot help us; we must look to kinetics for the answer. Both thermodynamic and kinetic aspects will be dealt with in this chapter.

EQUILIBRIUM

The equilibrium state is that state which leads to the greatest degree of randomness in the universe. All spontaneous processes lead to a net entropy increase for the system and its surroundings taken together. If the system undergoes a decrease in entropy in a spontaneous process, the surroundings must experience an even larger increase in entropy. Thus, for any spontaneous process

$$\Delta S_{univ} = \Delta S_{sys} + \Delta S_{sur} > 0$$

This inequality is a statement of the second law of thermodynamics. Although no formal proof of this law is possible, the absence of any experimental exceptions provides confidence in its validity.

Two examples of the use of the second law will best serve to demonstrate its applicability.

1. Flow of heat

Two identical copper blocks, one at temperature T_1 and the other at T_2, are placed in contact. Experience tells us that they will come to a common temperature $(T_1 + T_2)/2$. If no heat is allowed to pass between the blocks and the surroundings, the entropy change for the surroundings will be zero. Thus the entropy of the blocks must rise if the second law is valid. If the heat capacity is constant over the temperature range of interest, the entropy change for the first block will be

$$\Delta S_1 = \int_{T_1}^{(T_1+T_2)/2} \frac{C_V \, dT}{T}$$

$$= C_V \ln \frac{T_1 + T_2}{2T_1}$$

and for the second block,

$$\Delta S_2 = \int_{T_2}^{(T_1+T_2)/2} \frac{C_V \, dT}{T}$$

$$= C_V \ln \frac{T_1 + T_2}{2T_2}$$

The total entropy change will be

$$\Delta S_{\text{sys}} = C_V \ln \frac{(T_1 + T_2)^2}{4T_1 T_2}$$

If T_2 is equal to $T_1 + \Delta T$, then

$$\Delta S_{\text{sys}} = C_V \ln \frac{4T_1{}^2 + 4T_1 \Delta T + \Delta T^2}{4T_1{}^2 + 4T_1 \Delta T}$$

or

$$\Delta S_{\text{sys}} = C_V \ln \left(1 + \frac{\Delta T^2}{4T_1 T_2}\right)$$

Since the fraction $\Delta T^2/4T_1T_2$ cannot be negative, the total entropy change for the universe must be greater than zero, regardless of the sign of ΔT. The only case where the entropy change can be zero is that when $\Delta T = 0$. This corresponds to an ideal reversible process carried out at equilibrium; when $\Delta T = 0$, the blocks are in thermal equilibrium with each other before being brought together. For the spontaneous process, $\Delta T \neq 0$, the entropy of the universe must increase.

2. Mixing of gases

The contents of two identical flasks at the same temperature are allowed to mix spontaneously. One flask initially contains 1 mole of He and the other 1 mole of Ar. These gases behave nearly ideally, and so the presence of one will not affect the other. Each gas will "see" only that the volume of its container has been doubled. Thus the entropy change for this mixing process will be the same as the entropy change associated with doubling the volume of each gas separately at constant temperature:

$$\Delta S_{He} = \Delta S_{Ar} = \int_{V_1}^{V_2} \frac{P\, dV}{T} = \int_{V_1}^{V_2} \frac{R}{V}\, dV = R \ln \frac{V_2}{V_1} = R \ln 2$$

$$\Delta S_{sys} = \Delta S_{He} + \Delta S_{Ar} = 2R \ln 2$$

FREE ENERGY

In chemical studies our interest is focused on the system in which we are particularly interested. It would be more convenient in equilibrium considerations if we could direct our attention to only the properties of the system and define our criterion of equilibrium in terms of the system's properties only. Most important natural systems are subject to the restraint of constant pressure and temperature. Fortunately, such systems can be analyzed rather easily. The entropy of the surroundings can be affected only through exchange of heat with the system and hence is determined entirely by the heat balance of the system. As shown previously, the heat added to the system in a constant-pressure process is equal to its enthalpy increase, ΔH. As the surroundings constitute an infinite heat reservoir, the heat transfer may be considered reversible for the surroundings. Hence

$$\Delta S_{sur} = \frac{-\Delta H_{sys}}{T}$$

Thus for processes carried out at constant P and T the entropy change for the universe can be written wholly in terms of properties of the system:

$$\Delta S_{univ} = \Delta S_{sys} - \frac{\Delta H_{sys}}{T}$$

In order to put the criterion for equilibrium in terms of energy units rather than entropy units, chemists have rewritten the previous equation as

$$T \, \Delta S_{univ} = T \, \Delta S_{sys} - \Delta H_{sys}$$

A new property of the system, the Gibbs free energy, G, is then defined as

$$G = H - TS = E + PV - TS$$

and

$$\Delta G = \Delta H - T \, \Delta S - S \, \Delta T$$

For the processes carried out at constant temperature,

$$\Delta G = \Delta H - T \, \Delta S$$

and hence

$$\Delta G_{sys} = -T \, \Delta S_{univ}$$

Since ΔS_{univ} must increase for all natural processes, it is clear that ΔG_{sys} must decrease.   For isobaric, isothermal processes

$$\Delta G = \Delta H - T \, \Delta S$$

where all the parameters refer to the system.   Most of the remainder of this book will involve application of this equation to chemical systems in nature.   For any process to proceed spontaneously the system must satisfy the inequality $\Delta G < 0$.   If the system is at equilibrium, $\Delta G = 0$ for all possible processes.

Occasionally it is necessary to consider a system constrained to constant volume, but where the pressure may vary.   In these cases, the Gibbs free energy is not a particularly useful concept.   However, the entropy change of the universe may still be written in terms of properties of the system.   At constant volume the heat exchanged between the system and the surroundings is equal to ΔE_{sys} (the change in the internal energy of the system).   Thus

$$\Delta S_{sur} = \frac{-\Delta E_{sys}}{T}$$

The Helmholtz free energy, A, is then defined as $A = E - TS$. For isothermal processes, the change in the Helmholtz free energy of the system is given by

$$\Delta A = \Delta E - T \Delta S$$

If the process is carried out at constant volume,

$$\Delta A_{sys} = -T \Delta S_{univ}$$

a direct analogy to the constant-pressure case.

Although the property free energy† was initiated as a convenience, it has a direct physical significance. The quantity $-\Delta G$ represents the portion of the energy made available by any natural process which can be converted into useful work.

For example, during an isothermal chemical reaction carried out at constant pressure, an amount of heat, ΔH, is released. This heat normally flows into the surroundings, causing an increase in entropy. However, one might attempt to "capture" this energy and convert it into useful work. To perform this conversion quantitatively in cases where the entropy of the system *decreases* clearly would violate the second law of thermodynamics. If complete conversion of this heat into work were accomplished, the entropy change for the surroundings would be zero; hence the total entropy change would be that for the system alone. Thus the entropy of the universe would fall, violating the second law. Actually, regardless of the design of the system, if ΔS_{sys} is negative, an amount of heat equal to $-T \Delta S_{sys}$ must be used to increase the entropy of the surroundings. If the entropy of the system increases during a spontaneous process, an extra quantity of work is obtained over that derived from ΔH. The maximum energy, $-\Delta G$, made available by a spontaneous process for the operation of a heat engine or a battery is thus given by $T \Delta S - \Delta H$, the free energy released by the system.

An excellent example of this principle is provided by Carnot's ideal heat engine discussed in Chap. 4. It is easily shown that the fraction, f, of the heat energy extracted from the hot reservoir which can be converted to useful work is always less than unity. The net work done during a given cycle is the work performed on the surroundings during expansion minus the work done on the system during compression. Since the change in internal energy for a complete cycle must be zero, the net work done must equal the net gain of heat energy, which is the difference between the heat, q_H, added from the hot reservoir during the

---

† In all cases which follow, the term *free energy* will refer to Gibbs free energy unless otherwise stated.

isothermal expansion and that, q_C, lost to the cold reservoir during the compression.†

Hence

$$f = \frac{\text{net work done}}{\text{heat extracted}} = \frac{q_H - q_C}{q_H}$$

$$= 1 - \frac{q_C}{q_H}$$

However, since the entropy change per cycle must also be zero,

$$\frac{q_H}{T_H} = \frac{q_C}{T_C}$$

giving

$$f = 1 - \frac{T_C}{T_H} = \frac{T_H - T_C}{T_H}$$

Complete conversion of energy derived from the hot reservoir to useful work can be accomplished only if the cold reservoir is at absolute zero. For example, if T_H is 400°K and T_C is 300°K, f will be 0.25. Thus for each 100 units of heat energy removed from the hot reservoir, at least 75 must be lost to the cold reservoir. The useful work cannot exceed 25 units of energy.

This same result can be obtained by considering the entropy change associated with short-circuiting the same quantity of heat from the hot to the cold reservoir, with no attempt made to derive useful work. For such a process the entropy change for the hot reservoir is

$$\Delta S_H = \frac{-q}{T_H}$$

and for the cold reservoir,

$$\Delta S_C = \frac{+q}{T_C}$$

---

† All thermodynamic arguments can be formulated in terms of thermodynamic variables. In this case, q_H and q_C are equivalent to the enthalpy changes for the hot and cold reservoirs, respectively (the exchange of heat is carried out at constant temperature and pressure for the reservoirs). Since the classic treatment of Carnot cycles is always cast in terms of heat exchange, in order to acquaint the reader with the traditional mode of presentation we shall use heat exchange in this and the following examples.

If enthalpy change is used instead of the amount of heat exchanged, care must be taken that the enthalpy change of the reservoir and not of the gas is used. In an isothermal expansion the pressure of the gas obviously changes; however, enthalpy change can be equated to heat exchange only for constant-pressure processes. Thus q is not equal to ΔH_{gas} but is equal to $\Delta H_{\text{reservoir}}$, since the reservoir is maintained at constant pressure.

The net entropy change for the universe will be

$$\Delta S_{univ} = q\left(\frac{1}{T_C} - \frac{1}{T_H}\right) = q\frac{T_H - T_C}{T_C T_H} = \frac{q}{T_C}\frac{T_H - T_C}{T_H}$$

In terms of the free energy change for the cold reservoir,

$$\Delta G = -T_C \Delta S_{univ} = -q\frac{T_H - T_C}{T_H}$$

Further, since the enthalpy change for the cold reservoir, ΔH, is equal to q,

$$\Delta G = -\Delta H\frac{T_H - T_C}{T_H} = -f\Delta H$$

and

$$-\Delta G = f\Delta H$$

Again the negative of the change in free energy, $-\Delta G$, is that portion of the total heat energy made available by the system that can be used for useful work.

The change in free energy is also the amount of energy wasted during unharnessed natural processes. The energy is wasted in the sense that it generates more useless randomness rather than accomplishing useful work.

Now that we have shown that the most stable chemical configuration is that leading to the lowest free energy for the system, it is of interest to consider how this principle may be applied to chemical systems. The question most frequently asked is as follows: If a mixture of x moles of element A, y moles of element B, . . . , is confined at a temperature T_1 and a pressure P_1, in what chemical form will the atoms most stably coexist? If free energies (at T_1, P_1) were available for all conceivable combinations of the elements present, it would be possible to calculate the relative free energies for all possible combinations of these elements which would use up the available atoms and see which combination yielded the lowest free energy. In essence this is what the chemist does.

Two practical problems immediately arise in connection with such a procedure. First, absolute free energies cannot be measured for any element or compound. The only measurable quantities are the differences in free energy between chemical compounds. Second, tabulation of free energies for all compounds at all conceivable combinations of temperature and pressure would require tables of prohibitive bulk. The first problem is solved by assigning arbitrary absolute free energies to 92 compounds (one compound for each element). The compounds chosen are the most stable forms of each element under a given set of standard conditions (1 atm, 25°C). The free energy of elements under these conditions is taken to be zero. Whereas the job could have been done in innumerable ways, most tabulations of free energy data are based on this one method. In this system, diatomic oxygen gas, O_2, is assigned a free energy of zero at 1 atm, 25°C. Magnesium metal is assigned a free energy of zero under the same conditions. The free energy for the compound MgO would be equal to the change in free energy when $\frac{1}{2}$ mole of O_2 gas is combined with 1 mole of Mg metal:

$$Mg + \tfrac{1}{2}O_2 \rightarrow MgO$$

$$\Delta G = G_{MgO} - \tfrac{1}{2}G_{O_2} - G_{Mg}$$

Hence

$$G_{MgO} = \Delta G + \tfrac{1}{2}G_{O_2} + G_{Mg} = \Delta G$$

The same problem is encountered if one wishes to establish a table of enthalpies. Again, it is impossible to establish an absolute scale. The same logic is applied, and the enthalpy of any element in the standard state is assigned a value of zero. The enthalpy of any compound in the standard state is equal to the enthalpy change occurring when the compound is formed from its elements under standard conditions. This enthalpy change is called the *enthalpy of formation* or *heat of formation* of the compound. If heat is released to the surroundings when the compound is formed, the reaction is called *exothermic;* if heat is absorbed, the reaction is *endothermic.* Since the heat entering or leaving the surroundings is equal to $-\Delta H_{reac}$, an exothermic reaction has an enthalpy of reaction which is negative.

It should be emphasized that the absolute value of free energy or enthalpy assigned in this way has by itself no meaning. Differences between these free energies or enthalpies are, however, perfectly valid. Spontaneous chemical reactions will occur only when the free energy change is negative.

Once a list of the relative free energies of various compounds at the standard conditions is available, the next step is to extend these values to other sets of temperature and pressure conditions. Changes in free energy for any given compound can be computed provided its heat capacity at constant pressure, C_P, its coefficient of thermal expansion, α, and its compressibility, β, are known.

Let us first consider the pressure dependence of free energy. Since by definition

$$G = E + PV - TS$$

we have

$$\left(\frac{\partial G}{\partial P}\right)_T = \left(\frac{\partial E}{\partial P}\right)_T + P\left(\frac{\partial V}{\partial P}\right)_T + V - T\left(\frac{\partial S}{\partial P}\right)_T$$

However, as shown in Chap. 4,

$$dE = T\,dS - P\,dV$$

and

$$\left(\frac{\partial E}{\partial P}\right)_T = T\left(\frac{\partial S}{\partial P}\right)_T - P\left(\frac{\partial V}{\partial P}\right)_T$$

yielding

$$\left(\frac{\partial G}{\partial P}\right)_T = V$$

Hence, at constant temperature

$$dG = V\,dP$$

The free energy of any material rises with pressure in proportion to the volume it occupies. If, as is the case for solids, V undergoes only very small changes with pressure, then

$$G_{P_2} = G_{P_1} + V(P_2 - P_1)$$

If P_1 is the standard-state pressure, then for high pressures

$$P_2 - P_1 \cong P_2$$

giving

$$G_P \cong G_{std} + PV$$

This result explains why those forms which occupy the smallest volume eventually become stable as pressure rises.

For a gas, large changes in volume accompany pressure changes; hence

$$G_{P_2} - G_{P_1} = \int_{P_1}^{P_2} V\,dP$$

If the gas is ideal,

$$G_{P_2} - G_{P_1} = RT\int_{P_1}^{P_2} \frac{dP}{P} = RT\ln\frac{P_2}{P_1}$$

Taking P_1 to be 1 atm (i.e., standard conditions),

$$G_P = G_{std} + RT\ln P$$

For any substance the volume change with pressure can be incorporated into the calculation of free energy change with pressure if the compressibility of the substance is known:

$$V_P = V_{std} - \beta V_{std}\, dP = V_{std}(1 - \beta\, dP)$$

Thus, for the most accurate calculations, β must be known as a function of temperature and pressure.

The temperature dependence at constant pressure of free energy is also readily computed. Since

$$G = E + PV - TS$$

$$\left(\frac{\partial G}{\partial T}\right)_P = \left(\frac{\partial E}{\partial T}\right)_P + P\left(\frac{\partial V}{\partial T}\right)_P - T\left(\frac{\partial S}{\partial T}\right)_P - S$$

but

$$dE = T\, dS - P\, dV$$

Hence

$$\left(\frac{\partial E}{\partial T}\right)_P = T\left(\frac{\partial S}{\partial T}\right)_P - P\left(\frac{\partial V}{\partial T}\right)_P$$

and

$$\left(\frac{\partial G}{\partial T}\right)_P = -S$$

or at constant pressure

$$dG = -S\, dT$$

The entropy used in this calculation is the total chemical entropy accumulated by the substance between absolute zero and the temperature of interest, plus any entropy due to residual geometric disorder at absolute zero.

In cases where entropies based on heat-capacity integration disagree with those calculated from statistical-mechanical considerations, the difference can be attributed to a degree of residual geometric randomness in the crystal near absolute zero temperature. The difference between the statistically derived entropy (which is always the larger of the two) and the entropy based on heat capacity is called *residual entropy*. An example of a crystal which exhibits residual entropy is CO. The perfectly ordered crystal would appear as shown in Fig. 5.1a. The actual crystal is closer to that shown in Fig. 5.1b.

If a reaction at absolute zero involved a reactant with residual entropy, there *would* be an entropy change resulting from the reaction. It is for this reason that the third law specifies perfectly crystalline substances. If entropies are to be calculated from heat-capacity data for a compound which possesses residual entropy, this entropy must be added to the calculated value for use in the $dG = -S\,dT$ relationship.

Once entropy is known as a function of temperature, the variation of free energy with temperature can be calculated from the relationship

$$G_T = G_{\text{std}} - \int_{298}^{T} S\,dT$$

We can conclude from the preceding discussion that in order to predict the equilibrium configuration of a given system, we need the following data for the compounds formed by ingredients of the system:

1. Their volumes at standard conditions (1 atm, 25°C) and α and β as a function of T and P, which is the equation of state
2. C_P as a function of T and P
3. Their entropies at standard conditions (1 atm, 25°C)
4. Their free energies at standard conditions (1 atm, 25°C)

With this information the free energies of the various possible chemical forms of the system can be compared at any P and T. The assemblage yielding the lowest value for the total free energy is the stable form.

As was done for internal energy and enthalpy in Chap. 3 and for entropy in Chap. 4, the variation of the free energies of calcite and

```
C - O · · C - O · · C - O · · C - O · · C - O
  :       :       :       :       :
C - O · · C - O · · C - O · · C - O · · C - O
  :       :       :       :       :
C - O · · C - O · · C - O · · C - O · · C - O
  :       :       :       :       :
C - O · · C - O · · C - O · · C - O · · C - O
  :       :       :       :       :
C - O · · C - O · · C - O · · C - O · · C - O
                    (a)


O - C · · C - O · · C - O · · O - C · · O - C
  :       :       :       :       :
C - O · · O - C · · O - C · · C - O · · C - O
  :       :       :       :       :
C - O · · C - O · · O - C · · C - O · · O - C
  :       :       :       :       :
O - C · · C - O · · C - O · · O - C · · C - O
  :       :       :       :       :
C - O · · C - O · · C - O · · O - C · · O - C
                    (b)
```

Fig. 5.1. *(a)* Perfectly ordered crystal. *(b)* Crystal with residual entropy due to random orientation of atoms.

aragonite are calculated in Fig. 5.2.   This is easily done by subtracting $T \, \Delta S$ from ΔH.   The most interesting result of this calculation is that the sign of the free energy change is different at 5000 atm, 298°K than at the other three sets of conditions.   Aragonite replaces calcite as the stable form under these conditions.

The ingredients in chemical reactions can be placed in two categories: those whose free energy does not change during the course of the reaction and those whose free energy does change.   For example, during the reaction of calcite to form aragonite the molar free energies of calcite and of aragonite remain constant.   On the other hand, if O_2 gas and CO react to form CO_2 the molar free energies of the reactants fall as the reaction proceeds and that of CO_2 rises.

In the first case it is possible for the reactant (calcite) to disappear entirely.   In the second case neither O_2 nor CO_2 will ever entirely disappear.   Instead, a point will be reached when the free energy of the reactants has fallen to a level where it equals the rising free energy of the

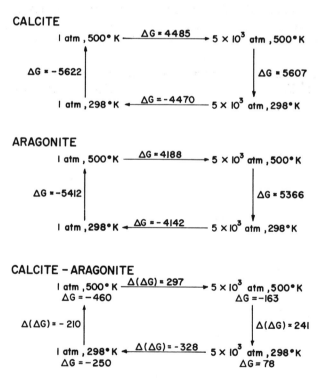

Fig. 5.2. Free energy cycle for calcite-aragonite.   All units are in calories.   The changes in ΔG refer to the reaction aragonite → calcite.   The free energies of formation at 1 atm and 298°K are calcite: $\Delta G_f° = -269{,}780$ cal;   aragonite: $\Delta G_f° = -269{,}530$ cal. At standard conditions, $\Delta G_{reac}° = -250$ cal.

product.   From this point on, the concentrations of O_2, CO, and CO_2 will
not change.   In problems involving reactants only of the first kind the
answer is of the "either-or" variety.   At equilibrium either compound A
will be present or it will be absent.

If one or more of the compounds involved changes its molar free
energy during the course of the reaction, then the answer must be given
as an equilibrium constant which expresses those conditions under which
the free energy change for the reaction becomes zero.   For the $2CO +
O_2 = 2CO_2$ reaction there is some fixed ratio of the square of the CO_2
partial pressure divided by the product of the O_2 partial pressure and the
square of the CO partial pressure for which equilibrium exists.   Any
combination of these three pressures yielding this constant will constitute
a stable mixture.   Therefore

$$K_p = \frac{p_{CO_2}{}^2}{p_{CO}{}^2 p_{O_2}}$$

The constant K will change with temperature.   Although at first glance
problems involving equilibrium constants do not appear to yield unique
solutions, as soon as the composition of the system is fixed (i.e., the number of C and O_2 atoms per unit volume) only one configuration can satisfy
both the equilibrium-constant and matter-conservation restrictions.

RATES

Thermodynamic equilibrium has not been established in many natural
systems.   Reaction rates are too slow to produce the thermodynamically
stable mixture in the time available.   If the predictions based on free
energy data are to be useful, we must have some criteria to judge under
what conditions rearrangement will go to completion.

Although reaction-rate theory is an extremely complex subject, one
very important generalization can be made.   Equilibrium is generally
established for processes carried out at high temperature and rarely for
processes carried out at low temperature.   The very strong temperature
dependence of reaction rates stems from the importance of energy barriers.   Two main steps are necessary for all reactions: (1) The reactants
must be brought into proximity and (2) they must collide with sufficient
vigor to combine.   In solids, for example, it is the mixing step which
generally limits the rate of reaction.   The reactants can move through
the crystal only by molecular diffusion.   In order to move from one
lattice site to another, a diffusing atom must shoulder aside the intervening
atoms.   It can do so only if it has a kinetic energy greater than some
minimum value.   An analogy is rolling a marble over a hill.   In order to

achieve this feat the marble must leave the valley with a kinetic energy at least as great as the gravitational energy corresponding to that released if the marble were dropped from the hilltop to the valley floor.   A marble with less than the critical energy would come to a halt and roll back before reaching the crest.   Similarly a diffusing atom is forced back to its initial position unless it has more than the so-called barrier energy.

If the reactants are rapidly mixed, as is the case in a gas, the reaction itself becomes the rate-limiting step.   In order to react, two colliding molecules must merge their outer electronic shells.   Because of the mutual repulsion between electrons, this merger requires an energetic collision.   In general, only a very small fraction of collisions between molecules are sufficiently energetic to result in reaction.

Thus, regardless of which step in the reaction is rate-limiting, the rate of reaction depends on the fraction of the molecules with energies in excess of that of some energy barrier.   As we shall see, this fraction rises rapidly with temperature; hence reaction rates rise rapidly with temperature.

The fraction of molecules with an energy exceeding the barrier energy can be calculated as follows: According to the Boltzmann distribution law, the fraction of the molecules in an energy level is given by

$$f_i = \frac{\exp(-E_i/RT)}{\Sigma \exp(-E_i/RT)}$$

where E_i is the energy of a mole of molecules in the given state.   If the energy states are closely spaced, then

$$df_i = \frac{-(1/RT)\exp(-E_i/RT)}{\Sigma \exp(-E_i/RT)}\, dE_i$$

Thus the fraction of molecules with energies between E and $E + dE$ is

$$df = \frac{(1/RT)\exp(-E/RT)\, dE}{q}$$

Integrating from the barrier energy to infinity, we get

$$f_{E_B} \to \infty = \frac{\int_{E_B}^{\infty} + (1/RT)\exp(-E/RT)\, dE}{q}$$

$$= \frac{-\exp(-E/RT)\Big|_{E_B}^{\infty}}{q}$$

$$= \frac{\exp(-E_B/RT)}{q}$$

Hence the fraction of the atoms with energies above the barrier energy, E_B, is proportional to $\exp(-E_B/RT)$.   If surmounting this barrier constitutes the rate-limiting step for the reaction, then the rate, A, must be given by

$$A = k \exp\left(-\frac{E_B}{RT}\right)$$

where k is the proportionality constant.   In logarithmic form it states

$$\ln A = \ln k - \frac{E_B}{R}\frac{1}{T}$$

Thus, when the log of the reaction rate is plotted against the reciprocal of the absolute temperature, a straight line whose slope is $-E_B/R$ should result.   This relationship was first proposed by Arrhenius in 1889.

When this relationship is used, the ratio of the reaction rate at temperature $T + \Delta T$ to the rate at temperature T is

$$\frac{A_{T+\Delta T}}{A_T} = \frac{\exp[-E_B/R(T+\Delta T)]}{\exp(-E_B/RT)} = \exp\left(\frac{E_B\,\Delta T}{RT^2}\right)$$

With a typical barrier energy of 18,000 cal/mole, an increase in temperature of 10 deg from 300 to 310°K would produce an increase in reaction rate given by

$$\frac{A_{310°K}}{A_{300°K}} = \exp\left(\frac{18,000}{300R}\frac{10}{300}\right) = e^1 = 2.76$$

A temperature increase of 200 deg would raise the reaction rate by a factor of e^{20}, an increase of about 1 billion times the 300°K rate.   Figure 5.3 shows a plot of the factor by which the reaction rate is increased per 10-deg temperature increase as a function of E_B.

In dealing with the reaction mechanisms, the role of catalysts must be considered.   A catalyst is a substance which participates in a reaction but is not consumed.   By combining in the presence of a catalyst, the reactants can merge more easily.   In other words, by providing an alternative path which has a lower activation energy, the catalyst increases the rate of the reaction.

Consider the example given above.   If the introduction of the catalyst were to reduce the barrier energy by a factor of 2, the increase in reaction rate would be

$$\frac{R_{9000}}{R_{18,000}} = \frac{\exp[-9000/(2\times300)]}{\exp[-18,000/(2\times300)]} = \exp\left(\frac{9000}{2\times300}\right) = e^{15} = 3.3\times10^6$$

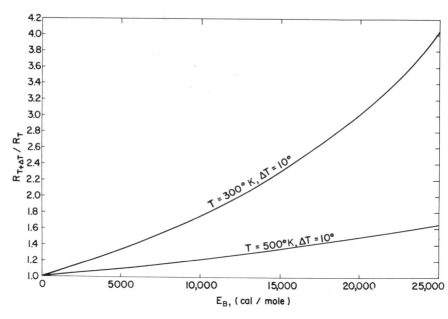

Fig. 5.3. $R(T + \Delta T)/R(T)$ as a function of barrier energy. $\Delta T = 10$ deg in both cases. The curves are labeled with the appropriate values of the initial temperature.

In natural systems, living organisms provide the most important catalysts. Using enzymes (organic catalysts) they are able to carry out reactions which normally would take place at negligible rates. With this concept of energy barriers in mind, we will consider two examples.

REACTION BETWEEN TWO GASES

First let us consider the reaction between two gas molecules X and Y to form a third molecule XY. It is assumed that the speed of the reaction is equal to the number of collisions per second between reacting molecules times the fraction of these collisions which are successful in producing a reaction.

The number of collisions per second in 1 cm³ of gas involving one molecule of X and one of Y can be shown to be

$$Z = \pi N_X N_Y \left(\frac{d_X + d_Y}{2} \right)^2 \sqrt{\frac{8kT}{\pi \mu}}$$

where N_X and N_Y are the number of molecules of X and Y per cubic centimeter, d_X and d_Y are their molecular diameters (in centimeters),

μ is their reduced mass $[m_X m_Y/(m_X + m_Y)]$ in grams, k is the Boltzmann constant, and T is the absolute temperature; the quantity $\sqrt{8kT/\pi\mu}$ is the average speed of molecules of type X relative to molecules of type Y. The mean collision cross section is given by

$$\pi \left(\frac{d_X + d_Y}{2} \right)^2$$

since $(d_X + d_Y)/2$ is the mean molecular diameter.   The collision rate is thus equal to the number of molecules of X times the number of molecules of Y times their relative average speed times the average collision cross section.   For most reactions Z is of the order of 10^{27} to 10^{28}.

Only a fraction of the collisions occurring in a gas are successful in producing a reaction.   This is because the energy associated with collisions depends on the kinetic energy of the molecules involved.   Since the molecular kinetic energies follow a Boltzmann distribution law, the collision energies must have a similar distribution.   In order to combine, the collision energy of X and Y must be at least E_B.   This requirement is shown schematically in Fig. 5.4.   The number of XY molecules formed per unit time, dN_{XY}/dt, is given by

$$\frac{dN_{XY}}{dt} = Z \exp\left(-\frac{E_B}{RT} \right)$$

This is also the rate of disappearance of X and Y.   This rate may be converted into units of moles per liter of substance reacting per second.

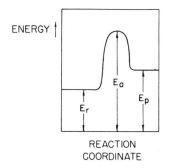

Fig. 5.4. Schematic diagram showing the energy changes taking place during a reaction. In order for the reactants to combine, they must collide with sufficient energy to overcome the barrier energy, $E_B = E_a - E_r$.

The number of moles per liter of X, C_X, is given by $(N_X/N_0)10^3$, where N_0 is Avogadro's number. Thus

$$\frac{dN_X}{dt} = \frac{N_0}{10^3}\frac{dC_X}{dt}$$

The rate law for our hypothetical reaction is

$$\frac{dC_X}{dt} = kC_XC_Y$$

Since

$$kC_XC_Y = k\frac{N_X}{N_0}\frac{N_Y}{N_0}10^6 = \frac{10^3}{N_0}Z\exp\left(-\frac{E_B}{RT}\right)$$

we can solve for k:

$$k = \frac{N_0}{N_XN_Y10^3}Z\exp\left(-\frac{E_B}{RT}\right)$$

Substituting for Z,

$$k = \frac{\pi N_0}{10^3}\left(\frac{d_X+d_Y}{2}\right)^2\sqrt{\frac{8kT}{\pi\mu}}\exp\left(-\frac{E_B}{RT}\right)$$

It is easily seen that, since the collision rate, Z, varies as the square root of T, its contribution to the increase of reaction rate with temperature is negligible.

REACTION TAKING PLACE WITHIN A SOLID

Although often exceedingly slow, the random diffusive mixing of atoms and molecules takes place in all materials. Where chemical gradients exist, this process can lead to net transport of constituents and hence can promote chemical reactions. Thus, where all other mechanisms of mixing fail, diffusion can always be called on. Its rate provides a minimum estimate for any process. Thus it is of considerable importance to understand how diffusion operates.

As convective motion almost always dominates molecular diffusion in liquids and gases, we will confine our discussion to solids. Just as the game of checkers would become impossible if all the squares were occupied, so diffusion would not take place in a perfect solid. Unoccupied positions must be available if movement is to be possible. All solids contain such vacant sites.   These so-called defects can arise in many ways.   Simple thermal agitation causes the volume of the solid to be slightly greater than the ideal volume by forcing atoms onto the surface.   This leads to more available sites than atoms, resulting in unoccupied positions.   Chemical substitution of a doubly charged cation for two singly charged cations leads to a hole.   The former is an example of a nonpermanent hole (all such holes would disappear at absolute zero) and the latter, of a permanent hole (their number remains constant with temperature).   Since the generation of a thermally induced hole requires a certain energy, ϵ_H, the number will be proportional to the number of atoms with energies more than ϵ_H.   Thus the total number of defects will be

$$N_{\text{tot}} = N_{\text{perm}} + A \exp\left(-\frac{\epsilon_H}{kT}\right)$$

or

$$N_{\text{tot}} = N_{\text{perm}} + A \exp\left(-\frac{E_H}{RT}\right)$$

Obviously the rate of diffusion will depend on the number of holes available.

The next problem is to define the probability that any given atom adjacent to a hole will make the jump from its lattice site into the vacant position.   The atom can move only if by chance it achieves a high enough energy to force its way between the intervening atoms.   The probability, p, of a successful jump is then given by the product of the number of attempts, B, per unit time by the atoms present (the vibration rate, C, times the number of holes) and the fractions of the atoms with a sufficiently high energy to shoulder aside the atoms which partially block the path to the hole:

$$p = B \exp\left(-\frac{\epsilon_B}{kT}\right) = B \exp\left(-\frac{E_B}{RT}\right)$$

Since B is given by

$$B = C \left[N_{\text{perm}} + A \exp\left(-\frac{E_H}{RT}\right) \right]$$

we get

$$p = CN_{\text{perm}} \exp\left(-\frac{E_B}{RT}\right) + CA \exp\left(-\frac{E_B + E_H}{RT}\right)$$

$$= m \exp\left(-\frac{E_B}{RT}\right) + n \exp\left(-\frac{E_B + E_H}{RT}\right)$$

At low temperatures the number of permanent holes dominates over the number of thermal holes, and movement rate, R', is given by

$$R' \cong m \exp\left(-\frac{E_B}{RT}\right)$$

At high temperature the reverse is true and

$$R' \cong n \exp\left(-\frac{E_B + E_H}{RT}\right)$$

Thus the rate of diffusion changes more rapidly with temperature in the high- than in the low-temperature range.

Diffusion rates are most commonly plotted as $\ln R'$ versus $1/T$, for in this form straight lines are obtained. For example,

$$R' = m \exp\left(-\frac{E_B}{RT}\right)$$

$$\ln R' = \ln m - \frac{E_B}{RT}$$

where $\ln m$ is the intercept at $1/T = 0$ and $-E_B/R$ is the slope of the straight line.   The relationship between the contribution of permanent and thermally induced holes to molecular movement is nicely shown by such a plot (see Fig. 5.5).

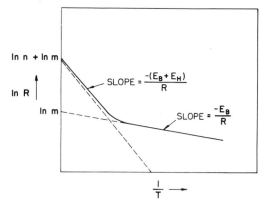

Fig. 5.5. Plot of the natural logarithm of diffusion rate versus $1/T$ for a hypothetical reaction.

Having established the rate at which molecules jump into holes, we must consider how such a random process can lead to a unidirectional transport.   In the absence of a concentration gradient no net movement can take place.   The random motions of the atoms lead only to internal mixing of the atoms present.   This so-called self-diffusion is of no consequence for problems involving chemical reactions.   A gradient is necessary if there is to be net transport by diffusion.

In the presence of a concentration gradient more atoms will jump into holes from the high- than from the low-concentration side, leading to a down gradient transport.   Intuitively, other factors remaining constant, the magnitude of the transport should be proportional to the magnitude of the gradient.   Hence Fick's law states that the flux, F, or grams of diffusing substance passing through a given cross section per unit time is proportional to the gradient dc/dx.   The proportionality constant D is called the diffusion constant for the medium; hence

$$F = -D \frac{dc}{dx}$$

From this law and some elementary geometric considerations it is possible to derive the following differential equation which relates the concentration of the diffusing substance to position and time.   In one dimension

$$\frac{\partial C}{\partial t} = D \frac{\partial^2 C}{\partial x^2}$$

As we shall see, given the initial distribution of the diffusing substance in the medium, its distribution at all subsequent times can be determined by using this equation (provided, of course, the appropriate diffusion constant is available).

The diffusion equation is derived as follows: Let us consider a cube of material such as shown in Fig. 5.6.   For diffusion along the X axis only, the total flux of material into the cube is given by

$$F_{\text{in}} = \left[f_{\text{C.P.}} - \left(\frac{\partial f}{\partial x}\right)_t \frac{dx}{2} \right] dy \, dz$$

Here $f_{\text{C.P.}}$ is the flux through the center plane, $(\partial f/\partial x)_t$ is the flux gradient at any given time, $-dx/2$ is the distance from the center plane to the edge of the cube, and $dy \, dz$ is the area of the face where the diffusing substance enters the cube.   The total flux of material out of the cube is

$$F_{out} = \left[f_{C.P.} + \left(\frac{\partial f}{\partial x} \right)_t \frac{dx}{2} \right] dy\; dz$$

The net gain of material by the cube is the change in mass with time,

$$\left(\frac{\partial m}{\partial t} \right)_x = F_{in} - F_{out}$$

$$= \left(\frac{\partial f}{\partial x} \right)_t \left[\left(-\frac{dx}{2} \right) - \left(\frac{dx}{2} \right) \right] dy\; dz$$

$$= - \left(\frac{\partial f}{\partial x} \right)_t dV$$

Since the change in mass per unit volume is the change in concentration,

$$\left(\frac{\partial}{\partial t} \frac{m}{dV} \right)_x = \left(\frac{\partial C}{\partial t} \right)_x = - \left(\frac{\partial f}{\partial x} \right)_t$$

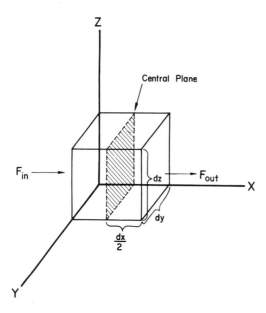

Fig. 5.6. Geometric model used to derive the diffusion law. See text for discussion.

But we know from Fick's law that

$$f = -D \left(\frac{\partial C}{\partial x}\right)_t$$

and so

$$\left(\frac{\partial f}{\partial x}\right)_t = -D \left(\frac{\partial^2 C}{\partial x^2}\right)_t$$

Thus

$$\left(\frac{\partial C}{\partial t}\right)_x = D \left(\frac{\partial^2 C}{\partial x^2}\right)_t$$

This means that the change in concentration of the diffusing species with time at any given point is proportional to the second derivative of concentration with respect to distance at the time of interest.

The diffusion coefficient, D, is proportional to the probability of an atom moving into a vacancy.   Hence

$$D = D_0 \exp\left(-\frac{E_D}{RT}\right)$$

where in the low-temperature range E_D is the barrier energy and in the high-temperature range is the sum of the barrier energy and the energy required to generate a hole.

No general solution can be given for the diffusion equation since the solution depends on the boundary conditions.   This difficulty can be overcome as follows: In the one-dimensional case (material homogeneous in the y and z directions, gradient along x only) the diffusing material can be treated as a large series of planar units each of width dx.   The diffusion of material away from each of these planes can be treated separately.   The concentration of diffusing material at any given distance, x, and time, t, can then be obtained by summing the contributions of material diffused to x from each of the planar units.

Let us first consider the distribution of material diffusing away from one such plane.   If the coordinate of the plane is taken to be zero, the concentration of diffusing material as a function of x and t will be

$$C_{xt} = \frac{b}{\sqrt{t}} \exp\left(-\frac{x^2}{4Dt}\right)$$

That this is a solution of the differential equation can be shown as follows:

Taking the derivative of the solution with respect to t,

$$\left(\frac{\partial C}{\partial t}\right)_x = -\frac{1}{2}\frac{b}{t^{\frac{3}{2}}}\exp\left(-\frac{x^2}{4Dt}\right) + \frac{bx^2}{4Dt^{\frac{5}{2}}}\exp\left(-\frac{x^2}{4Dt}\right)$$

and the first and second derivatives with respect to x,

$$\left(\frac{\partial C}{\partial x}\right)_t = \frac{b}{\sqrt{t}}\frac{-2x}{4Dt}\exp\left(-\frac{x^2}{4Dt}\right)$$

and

$$\left(\frac{\partial^2 C}{\partial x^2}\right)_t = -\frac{1}{2}\frac{b}{Dt^{\frac{3}{2}}}\exp\left(-\frac{x^2}{4Dt}\right) + \frac{bx^2}{4D^2t^{\frac{5}{2}}}\exp\left(-\frac{x^2}{4Dt}\right)$$

Hence

$$\left(\frac{\partial C}{\partial t}\right)_x = D\left(\frac{\partial^2 C}{\partial x^2}\right)_t$$

proving that our solution is valid.

The constant of integration, b, can be evaluated by noting that

$$M = \int_{-\infty}^{\infty} C\, dx$$

which states that the sum of all the diffused material must equal the amount, M, on the plane initially. Substituting for C,

$$M = \int_{-\infty}^{\infty}\frac{b}{\sqrt{t}}\exp\left(-\frac{x^2}{4Dt}\right)dx = 2b\sqrt{\pi D}$$

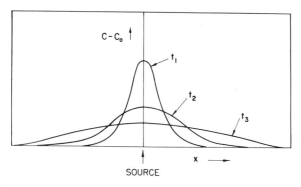

Fig. 5.7. Schematic diffusion diagram showing the concentration of a substance diffusing from a plane as a function of distance from the plane for various times after the start of diffusion, $t_1 < t_2 < t_3$.

yielding

$$b = \frac{M}{2\sqrt{\pi D}}$$

and

$$C = \frac{M}{2\sqrt{\pi Dt}} \exp\left(-\frac{x^2}{4Dt}\right)$$

Figure 5.7 shows schematically the distribution of concentration for various times after the onset of diffusion, and Fig. 5.8 shows the time dependence of the concentration at any given distance. The concentration of diffusing material at any given distance will initially rise, reach a maximum, and then begin to fall. The time required for the maximum to be achieved is a useful index of the rate at which diffusion proceeds. At the maximum

$$\left(\frac{\partial C}{\partial t}\right)_x = 0$$

$$\frac{M}{2\sqrt{\pi D}} \exp\left(-\frac{x^2}{4Dt}\right)\left(\frac{t^{-\frac{5}{2}}x^2}{4D} - \frac{t^{-\frac{3}{2}}}{2}\right) = 0$$

This is true when

$$t = \frac{x^2}{2D}$$

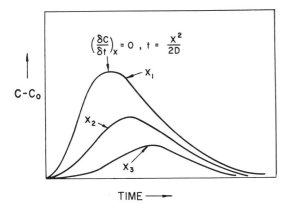

TIME ⟶

Fig. 5.8. Schematic diffusion diagram showing the concentration of a substance diffusing from a plane as a function of time at various fixed distances from the plane, $x_1 < x_2 < x_3$.

Hence the time required for the maximum to be reached is proportional to the square of the distance and inversely proportional to the value of the diffusion constant. If at room temperature $D = 1 \times 10^{-23}$ cm^2/sec, then 10^{11} years will be required for the maximum to be achieved at a distance of $\frac{1}{10}$ mm and 10^{15} years for a distance of 1 cm. If the value of E_D is 20,000 cal/mole, the diffusion rate will rise initially by the factor given in Fig. 5.3 for each 10°C. As the temperature becomes higher, the gain in diffusion rate for a 10-deg temperature increase becomes smaller. At 425°C, 10^7 years would pass before the concentration reached its maximum at 1 cm. At 625°C, the time would be 5×10^5 years.

Beyond this elementary treatment, the ramifications of diffusion problems become endless. Solutions of the diffusion equation have been worked out for many different geometries. For some problems the appropriate geometry is a sphere; for others, cylinders are useful; for still others, the infinite-slab solution is applicable. Fortunately, the solutions for different geometries and initial conditions are available in specialized books.

RATE LAWS AND REACTION ORDER

Experimental studies of reaction rates all take the same general form. The reaction is carried out in the laboratory at a fixed temperature and pressure. A known quantity of reactants is mixed, and the concentrations of reactants and/or products are measured (by following the change in some property of the system) at appropriate intervals of time. This gives the basic data of rate studies, the change in concentration of a given species with time. The experiment is then repeated at a series of different temperatures and pressures. From these data one attempts to determine the *order* of the reaction.

The order of a reaction is defined as the sum of the exponents in the *rate law*. For the reaction A + B → C the rate law would take the form

$$\frac{dC}{dt} = k_P[A - C]^m[B - C]^n$$

This equation states that the increase in concentration of C with time (dC/dt) is equal to a constant, k_P, times the instantaneous concentration of A (which is A − C) raised to the mth power times the instantaneous concentration of B (given by B − C) raised to the nth power. The subscript P generally indicates the order of the reaction, which is $m + n$. The constants, m and n, may be any number, including fractions. A negative exponent indicates that a substance retards the reaction.   The exponent may also be zero; in this case the reaction is independent of the concentration of that species.   It is important to realize that the stoichiometric equation for a reaction generally gives no clue to its rate law. The rate-law exponents may be the same as the stoichiometric coefficients, but this is by no means necessarily, or even generally, true. The only case where rate-law exponents may be deduced from stoichiometry is for reactions which take place in one step.   This would be true for a gaseous reaction which occurred as the result of a single collision.

A few examples will help to clarify the concepts of reaction order and rate laws.   A process of considerable significance in geologic systems is the spontaneous decay of naturally occurring radioactive isotopes.   The decay of any radioactive nuclide is a *first-order* process.   It depends only on the amount of the radioactive substance present.   For the decay of U^{238} to Th^{234}, the rate law is

$$- \frac{dN_{U^{238}}}{dt} = \lambda N_{U^{238}}$$

where N is the number of U^{238} nuclei present and λ is the decay constant (i.e., rate constant) for U^{238} decay.   Since the first power of N appears in the equation, the reaction is first order.

The decomposition of NO_2 follows the reaction

$$2NO_2 \rightarrow 2NO + O_2$$

The experimental rate law is

$$- \frac{d[NO_2]}{dt} = k_2[NO_2]^2$$

Thus the reaction is second order.

The photodecomposition of O_3 is an important process in shielding the earth from harmful ultraviolet radiation.   The reaction is

$$2O_3 \xrightarrow{h\nu} 3O_2$$

The experimental rate law is

$$- \frac{d[O_3]}{dt} = k \frac{[O_3]^2}{[O_2]}$$

This reaction is second order with respect to O_3 and minus first order with respect to O_2. Thus the presence of O_2 retards the decomposition of O_3. This example illustrates well the pitfalls of relating rate-law exponents to stoichiometry when complicated reaction mechanisms are involved.

The reaction order is controlled totally by the mechanism or pathway of the reaction. In the case of radioactive decay, the parent nuclide is unstable and emits a particle transforming itself into another nuclide. The reaction mechanism involves only the parent nuclide, and the reaction is therefore first order. In the case of the simple gas reaction

$$A + B \rightarrow C$$

when C is stable and does not inhibit the reaction, a two-body collision between A and B is required to produce a reaction. Thus the mechanism depends on both A and B, and the reaction is second order.

For most purposes it is necessary to have an integrated form of the rate law. For radioactive decay the rate law is

$$-\frac{dN}{dt} = \lambda N$$

In order to integrate this equation we must separate the variables:

$$-\frac{dN}{N} = \lambda \, dt$$

Performing the indefinite integral, since λ is not a function of time,

$$\ln N = -\lambda t + C$$

We may evaluate the constant C by examining this equation at the starting time, $t = 0$. If the initial number of atoms present is N_0, then

$$\ln N_0 = C$$

Thus

$$\ln N - \ln N_0 = -\lambda t$$

$$\ln \frac{N}{N_0} = -\lambda t$$

$$N = N_0 \exp\,(-\lambda t)$$

The half-life $(t_{\frac{1}{2}})$ of any first-order reaction is defined as the time at which the concentration of the rate-determining species reaches one-half of its initial value.   When

$$\frac{N}{N_0} = \frac{1}{2}$$
$$\ln \tfrac{1}{2} = -\lambda t_{\frac{1}{2}}$$
$$\ln 2 = \lambda t_{\frac{1}{2}}$$

or

$$t_{\frac{1}{2}} = \frac{0.693}{\lambda}$$

For U^{238}, $\lambda = 1.54 \times 10^{-10}$ year^{-1} and $t_{\frac{1}{2}} = 4.51 \times 10^9$ years.

Similar equations hold for any first-order process. If we have experimental rate data for a chemical reaction and wish to test whether the reaction is first order, we would plot the logarithm of the instantaneous concentration of the reactive species versus time. If this plot gives a straight line of negative slope, the reaction is first order with respect to that species.

Second- and higher-order rate laws may be integrated in an analogous fashion; however, the mathematics becomes considerably more cumbersome. Once the integrated rate law is obtained, the appropriate concentration expression can be plotted against time to determine the order of any reaction.

PROBLEMS

5.1 One mole of Ar at $T = 300°K$, $P = 1$ atm is mixed with 2 moles of Ne at $T = 400°K$, $P = 1$ atm. No heat is allowed to escape to the surroundings. If the gases are assumed to be ideal, what are the entropy and free energy changes for this process?

5.2 Using the data given in Fig. 5.2 and Table 3.6, calculate the free energy change for the reaction calcite → aragonite at $T = 400°K$, $P = 1$ kbar and for $T = 400°K$, $P = 10$ kbars.

5.3 Hanson and Gast have shown that Ar diffuses from biotite with an activation energy of approximately 50 kcal/mole for volume diffusion. If $D_0 = 1 \times 10^6$ cm^2/sec, calculate D for $T = 100, 200, 300$, and $400°C$.

5.4 A zircon contains 100 ppm of U ($U^{238}/U^{235} = 137.8$) and 100 ppm of Pb^{206}. No Pb^{204} is present. What is the age of the zircon? How would your answer change if 1 ppm of Pb^{204} were present?

*5.5 A globigerina ooze with uniform sedimentation rate has a C^{14}/C^{12} ratio at a depth of 30 cm equal to 0.125 that at its top. If $Th^{230}/U^{238} = 44$, $Ra^{226}/U^{238} = 0.25$, and $U^{234}/U^{238} = 1.00$ (all in terms of activity) at the top of the core, what will these ratios be at a depth of 150 cm?

5.6 A simplified model of gas exchange between air and water assumes that the main barrier to gas transfer is a stagnant water film at the interface through which gases pass only by molecular diffusion. The air above and the water below this film are assumed to be uniformly mixed. The thickness of the film is inversely related to the wind stress on the water surface.

Let us consider the oxygen balance in a stream of mean depth 100 cm. In this stream the consumption of O_2 by animal life exceeds the production by plants so that a net consumption of 10 moles of O_2 per cubic meter per year occurs. If the wind stress on this stream is such that the boundary layer is 50 microns thick, what is the degree of O_2 saturation ($[O_2]$ stream/αp_{O_2} air) at steady state? The solubility of O_2 in the stream water is 1.4 moles/m^3-atm, and the partial pressure of O_2 in the air is 0.2 atm. The diffusion constant for O_2 in water is 1×10^{-5} cm^2/sec at the stream temperature.

5.7 Strontium 90, a product of nuclear testing, has been largely transferred by rainfall from the atmosphere to the surface of the earth. That portion reaching the oceans remains in dissolved form. If vertical mixing in the surface oceans is accomplished by the diffusion of turbulent eddies, it should be possible to determine the appropriate diffusion constant from the extent to which vertical mixing of Sr^{90} has taken place. If as of 1970 it is found that the mean depth of penetration is 200 m, taking the mean deposition date to be 1960, what is the diffusion constant? Sr^{90} has a half-life of 30 years for radioactive decay. How will the decrease in abundance of Sr^{90} in the upper 200 m of surface water between 1970 and 2000, resulting from downward mixing, compare with the decrease due to radioactive decay during this time?

SUPPLEMENTARY READING

Anderson, T. F.: Self-diffusion of Carbon and Oxygen in Calcite by Isotope Exchange with Carbon Dioxide, *J. Geophys. Res.*, **74**: 3918 (1969).
Bischoff, J. L.: Kinetics of Calcite Nucleation: Magnesium Ion Inhibition and Ionic Strength Catalysis, *J. Geophys. Res.*, **73**: 3315 (1968).
Crank, J.: "The Mathematics of Diffusion," Oxford University Press, Fair Lawn, N.J., 1956.

Evernden, J. F., G. H. Curtis, R. W. Kistler, and J. Obradovich: Argon Diffusion in Glauconite, Microcline, Sanidine, Leucite and Phlogopite, *Am. J. Sci.*, **258**: 583 (1960).

Hanson, G. N., and P. W. Gast: Kinetic Studies in Contact Metamorphic Zones, *Geochim. Cosmochim. Acta*, **31**: 1119 (1967).

Jaeger, J. C.: Thermal Effects of Intrusions, *Rev. Geophys.*, **2**: 443 (1964).

Wollast, R.: Kinetics of the Alteration of K-feldspar in Buffered Solutions at Low Temperature, *Geochim. Cosmochim. Acta*, **31**: 635 (1967).

chapter six Reactions in Natural Gases

Equilibria involving gases are important in many natural systems. The compositions of the planetary atmospheres depend on the proportions of the chemical elements present and on the reactions which they undergo. In magmatic processes a gaseous phase is often important, particularly in the late stages of differentiated intrusions. In metamorphic processes a gaseous phase may provide the medium for transporting chemical species.

Reactions among gases could be discussed entirely in terms of the Gibbs free energy. However, it is useful to use instead the concept of chemical potential. For a pure substance the standard chemical potential, $\mu^\circ(298)$, is the Gibbs free energy at standard conditions (1 atm, $298^\circ K$) per mole of substance. We will first discuss the properties of the chemical potential and other partial molal variables and then use the concept of chemical potential to discuss equilibria in ideal gaseous mixtures. Finally, we will consider the problems of treating equilibria involving real gases which do not follow the ideal gas law.

THE CHEMICAL POTENTIAL

The chemical potential, μ, of a pure substance is defined as the change in free energy of the substance as the number of moles, n, of substance is changed.   Thus

$$\mu \equiv \left(\frac{\partial G}{\partial n}\right)_{T,P}$$

For a pure substance, μ is merely the Gibbs free energy per mole. This definition may seem unnecessary.   However, it allows us to extend the concept of free energy to mixtures.   For substance i in a multicomponent system the chemical potential, μ_i, is given by

$$\mu_i = \left(\frac{\partial G}{\partial n_i}\right)_{T,P,n_j}$$

where G is the total free energy of the mixture, n_i is the number of moles of substance i, and the subscript n_j indicates that the amount of all other substances in the system remains constant.

In addition to chemical potential, a whole group of special quantities are defined for problems involving mixtures.   These variables are called *partial molal* quantities.   For example, the partial molal volume of substance i in a mixture is defined by

$$\bar{V}_i \equiv \left(\frac{\partial V}{\partial n_i}\right)_{T,P,n_j}$$

In general, for any variable X, the partial molal X of species i is

$$\bar{X}_i = \left(\frac{\partial X}{\partial n_i}\right)_{T,P,n_j}$$

Partial molal quantities have the convenient property that they may be summed to give the total value of the variable for the system.   Thus the total volume, V, is equal to the sum of the partial molal volumes multiplied by the appropriate number of moles of each substance:

$$V = \Sigma n_i \bar{V}_i$$

The partial molal volume of such a substance is also the response of the *total* system to a change in the amount of this substance alone.   It should be noted that the partial molal volume need not bear any direct relationship to the molar volume of the pure substance.   In some cases, the partial molal volume of real gases may actually be negative; however, for the special case of ideal gases the partial molal volume is identical to the molar volume.

The chemical potential was defined in such a way that it is also the partial molal free energy. This allows us to write the variation in total free energy of a mixture as

$$dG = -S\,dT + V\,dP + \Sigma\mu_i\,dn_i$$

It is a theorem of calculus that crossed partial derivatives are equal. We may thus find the variation of the chemical potential with temperature and pressure:

$$\left(\frac{\partial\mu_i}{\partial P}\right)_{T,n_i,n_j} = \left(\frac{\partial V}{\partial n_i}\right)_{T,P,n_j} \equiv \bar{V}_i$$

$$\left(\frac{\partial\mu_i}{\partial T}\right)_{P,n_i,n_j} = -\left(\frac{\partial S}{\partial n_i}\right)_{T,P,n_j} \equiv -\bar{S}_i$$

It is interesting to note that the variation in chemical potential for a single member of a mixture (despite any nonideality) is analogous to the variation of free energy for a pure substance.

EQUILIBRIA INVOLVING IDEAL GASES

We have shown in Chap. 5 that the free energy of a substance at any temperature and pressure can be expressed in terms of a standard free energy plus terms representing the variation of free energy with temperature and pressure:

$$G(T,P) = G°(1\text{ atm}, 298°\text{K}) - \int_{298}^{T} S\,dT + \int_{1}^{P} V\,dP$$

For an ideal gas the last term becomes

$$\int_{1}^{P} V\,dP = RT \ln P$$

The chemical potential of a pure substance is its Gibbs free energy per mole. Hence

$$\mu(T,P) = G°(1\text{ atm}, 298°\text{K}) - \int_{298}^{T} S\,dT + RT \ln P$$

The first two terms on the right-hand side of this equation are combined to give a standard chemical potential, $\mu°(T)$. It should be noted that the standard chemical potential is a function of temperature, whereas the standard free energy is defined for a single specific temperature. The final result for the chemical potential of an ideal gas is

$$\mu(T,P) = \mu°(T) + RT \ln P$$

Before a chemical reaction can occur, the reactants must become intimately mixed. For this reason we must concern ourselves with gaseous mixtures. Ideal gases may be components in ideal gaseous mixtures. Each gas in the mixture will be characterized by a partial pressure, p_i. Since ideal gases undergo no intermolecular collisions, each gas will exert a pressure which will not depend on the other gases in the mixture. We may thus use the ideal gas law for each component separately:

$$p_i = n_i \frac{RT}{V}$$

where n_i is the number of moles of i and V is the total volume. For the whole system

$$P = N \frac{RT}{V}$$

where P is the total pressure and N the total number of moles of gas. Combining the above two relations,

$$p_i = \frac{n_i}{N} P$$

The quantity n_i/N is the mole fraction, X_i, of substance i in the mixture:

$$p_i = X_i P$$

For each gas in the mixture

$$\mu_i = \mu_i° + RT \ln p_i$$

and hence

$$\mu_i = \mu_i° + RT \ln X_i + RT \ln P$$

Since $\mu_i°$ is a function only of temperature (and hence independent of composition), we may evaluate μ_i° when $X_i = 1$. Under these conditions i is a pure gas, and μ_i° must be μ° for the pure gas.

We now have the necessary tools to solve problems involving mixtures of ideal gases. An example will best illustrate the nature of such problems. Consider the reaction of nitrogen with hydrogen to give ammonia:

$$N_2 + 3H_2 \rightleftharpoons 2NH_3$$

The double arrows indicate that the reaction may proceed in either direction. The total free energy of the system is

$$G = n_{N_2}\mu_{N_2} + n_{H_2}\mu_{H_2} + n_{NH_3}\mu_{NH_3}$$

Using the expression above for μ,

$$G = n_{N_2}\mu_{N_2}^\circ + n_{H_2}\mu_{H_2}^\circ + n_{NH_3}\mu_{NH_3}^\circ + RT(n_{N_2} + n_{H_2} + n_{NH_3}) \ln P$$
$$+ RT(n_{N_2} \ln X_{N_2} + n_{H_2} \ln X_{H_2} + n_{NH_3} \ln X_{NH_3})$$

Let us assume that the reaction was started by mixing some H_2 gas with N_2 gas. The number of moles of NH_3 present at any later time will be equal to twice the number of moles of N_2 which have disappeared. We can then calculate the total free energy of any mixture of the three gases. The minimum value found for G will correspond to the equilibrium concentrations of the three gases. We could carry out this calculation for any initial amounts of H_2 and N_2, but to simplify the arithmetic we will assume that

$$n_{H_2}^\circ = 3 \text{ moles}$$
$$n_{N_2}^\circ = 1 \text{ mole}$$

in the initial mixture. Thus

$$n_{H_2} = 3n_{N_2}$$
$$n_{NH_3} = 2(1 - n_{N_2})$$

at any time, since for every N_2 molecule which disappears, two NH_3 molecules are formed and three H_2 molecules disappear. Substituting in the expression for G,

$$G = n_{N_2}\mu_{N_2}^\circ + 3n_{N_2}\mu_{H_2}^\circ + 2(1 - n_{N_2})\mu_{NH_3}^\circ$$
$$+ RT[n_{N_2} + 3n_{N_2} + 2(1 - n_{N_2})] \ln P$$
$$+ RT[n_{N_2} \ln X_{N_2} + 3n_{N_2} \ln X_{H_2} + 2(1 - n_{N_2}) \ln X_{NH_3}]$$

The mole fractions are

$$X_{N_2} = \frac{n_{N_2}}{n_{N_2} + 3n_{N_2} + 2(1 - n_{N_2})} = \frac{n_{N_2}}{2(1 + n_{N_2})}$$

$$X_{H_2} = \frac{3n_{N_2}}{2(1 + n_{N_2})}$$

$$X_{NH_3} = \frac{1 - n_{N_2}}{1 + n_{N_2}}$$

Thus

$$G = n_{N_2}(\mu^{\circ}_{N_2} + 3\mu^{\circ}_{H_2} - 2\mu^{\circ}_{NH_3}) + 2\mu^{\circ}_{NH_3} + RT(2 + 2n_{N_2}) \ln P$$

$$+ RT \left[n_{N_2} \ln \frac{n_{N_2}}{2(1 + n_{N_2})} + 3n_{N_2} \ln \frac{3n_{N_2}}{2(1 + n_{N_2})} \right.$$

$$\left. + 2(1 - n_{N_2}) \ln \frac{1 - n_{N_2}}{1 + n_{N_2}} \right]$$

Setting the total pressure $P = 1$ atm and combining terms in the final bracket, we have

$$G = n_{N_2}(\mu^{\circ}_{N_2} + 3\mu^{\circ}_{H_2} - 2\mu^{\circ}_{NH_3}) + 2\mu^{\circ}_{NH_3}$$

$$+ RT[3n_{N_2} \ln 3 + 4n_{N_2} \ln n_{N_2} - 2(1 + n_{N_2}) \ln 2(1 + n_{N_2})$$

$$+ 2(1 - n_{N_2}) \ln 2(1 - n_{N_2})]$$

The terms each have a definite physical meaning. Thus $2\mu^{\circ}_{NH_3}$ is the free energy of a system consisting only of NH_3. The term $n_{N_2}(\mu^{\circ}_{N_2} + 3\mu^{\circ}_{H_2} - 2\mu^{\circ}_{NH_3})$ is equal to $-\Delta G^{\circ}_T$ for the reaction. The final term

$$RT \left[n_{N_2} \ln \frac{n_{N_2}}{2(1 + n_{N_2})} + 3n_{N_2} \ln \frac{3n_{N_2}}{2(1 + n_{N_2})} + 2(1 - n_{N_2}) \ln \frac{1 - n_{N_2}}{1 + n_{N_2}} \right]$$

is the free energy of mixing the three gases together. This term is zero only if pure NH_3 is present without any N_2 or H_2. For all other cases this term is negative. This is because entropy increases when the gases are mixed, and thus free energy decreases. It is the free energy of mixing which causes a mixture of three gases to have a lower free energy than the product NH_3 alone.

Figure 6.1 shows the variation of G as the reaction proceeds. The free energy of the system is a minimum when $n_{N_2} = 0.03$ mole, $n_{H_2} = 0.09$ mole, and $n_{NH_3} = 1.94$ moles.

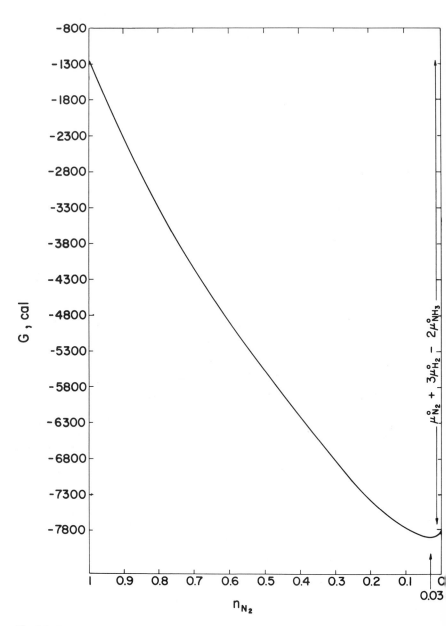

Fig. 6.1. The free energy (G) of a reacting mixture of N_2, H_2, and NH_3 plotted as a function of composition. The final mixture will have the composition indicated by the minimum on the curve.

Any problem involving gaseous equilibria could be handled by the method used above; however, this method is time-consuming, and the arithmetic becomes unmanageable in complicated reactions. Fortunately, there is an easier way to solve the problem. If we had tried several different starting mixtures of $N_2 + H_2$ or $N_2 + NH_3$ or $H_2 + NH_3$ we would have discovered that, provided the temperature remained the same, all the equilibrium mixtures shared a common feature. That is,

$$\frac{p_{NH_3}{}^2}{p_{N_2}p_{H_2}{}^3} = \text{const}$$

This constant is called the *equilibrium constant*, K_p. The subscript p denotes that it is a ratio of partial pressures. For the reaction $N_2 + 3H_2 \rightleftharpoons 2NH_3$, $K_p = 6.17 \times 10^6$ atm^{-2} at 298°K.

We may derive the general expression for the equilibrium constant by using the properties of an ideal gaseous mixture. We have stated above that the chemical potential of each member of such a mixture is given by

$$\mu_i = \mu_i^\circ + RT \ln p_i$$

Let ν_i = the coefficient of i in the balanced chemical reaction, with a $+$ sign for products and a $-$ sign for reactants, for example, $\nu_{NH_3} = 2$ for the reaction $N_2 + 3H_2 \rightleftharpoons 2NH_3$ and $\nu_{H_2} = -3$ for the same reaction. The change in chemical potential as the reaction proceeds is

$$\Sigma\mu(\text{products}) - \Sigma\mu(\text{reactants}) = \Sigma\nu_i\mu_i$$

$$\Sigma\nu_i\mu_i = \Sigma\nu_i\mu_i^\circ + RT\Sigma\nu_i \ln p_i$$

Two properties of logarithms allow us to simplify this expression. First,

$$\nu_i \ln p_i = \ln p_i{}^{\nu_i}$$

Second,

$$\Sigma \ln p_i{}^{\nu_i} = \ln \Pi p_i{}^{\nu_i}$$

(The symbol Π indicates that the $p_i{}^{\nu_i}$ are multiplied together.) Thus

$$\Sigma\nu_i\mu_i = \Sigma\nu_i\mu_i^\circ + RT \ln \Pi p_i{}^{\nu_i}$$

Since at equilibrium the free energy change becomes zero,

$$\Sigma\nu_i\mu_i = 0 \qquad \text{at equilibrium}$$

Thus

$$\Sigma\nu_i\mu_i^\circ = -RT \ln \Pi p_i{}^{\nu_i}$$

For the NH_3 reaction we have

$$2\mu^o_{NH_3} - \mu^o_{N_2} - 3\mu^o_{H_2} = -RT \ln \frac{p_{NH_3}{}^2}{p_{H_2}{}^3 p_{N_2}}$$

The term $\Sigma \nu_i \mu_i^o$ is simply ΔG^o for the reaction carried out at temperature T. The product of partial pressures is defined to be the equilibrium constant, K_p. Hence

$$\Delta G^o = -RT \ln K_p$$

and

$$K_p = \exp\left(-\frac{\Delta G^o}{RT}\right)$$

This expression allows us to calculate equilibrium constants from tabulated values of the free energy of reactants and products.

Since μ^o is a function only of temperature, K_p must also be a function only of temperature and is independent of the total pressure of the system. (This statement is strictly true only for gases which follow the ideal gas law.)

The temperature dependence of K_p can be found as follows:

$$G = H - TS$$

But

$$S = -\left(\frac{\partial G}{\partial T}\right)_{P,n_i}$$

Thus

$$G = H + T\left(\frac{\partial G}{\partial T}\right)_{P,n_i}$$

and

$$\frac{H}{T^2} = \frac{G}{T^2} - \frac{1}{T}\left(\frac{\partial G}{\partial T}\right)_{P,n_i}$$

But

$$\frac{\partial(G/T)}{\partial T} = \frac{1}{T}\frac{\partial G}{\partial T} - \frac{G}{T^2}$$

Thus

$$\frac{\partial(G/T)}{\partial T} = -\frac{H}{T^2}$$

The change in $\ln K_p$ with temperature is

$$\frac{d \ln K_p}{dT} = -\frac{1}{R}\frac{d(\Delta G^o/T)}{dT}$$

Hence

$$\frac{d \ln K_p}{dT} = \frac{\Delta H^\circ}{RT^2}$$

This equation may also be written

$$\frac{d \ln K_p}{d(1/T)} = -\frac{\Delta H^\circ}{R}$$

If the $\ln K_p$ is plotted against $1/T$, a straight line with slope $-\Delta H^\circ/R$ results. For reactions where the concentrations of species present at equilibrium can be measured, this method is used to determine experimentally the enthalpy of reaction.

EQUILIBRIUM INVOLVING MANY GASES

The gases H_2, H_2O, CH_4(methane), CO, CO_2, N_2, and NH_3 are present in many planetary atmospheres. An understanding of the equilibria involving these gases will aid us in investigating atmospheric properties. Many chemical reactions may be written relating the gases listed above; however, if it is assumed no other compounds can form, only three of these equations will be linearly independent. One set of three equations is

(1) $$4H_2 + CO_2 \rightleftharpoons CH_4 + 2H_2O$$

(2) $$H_2 + CO_2 \rightleftharpoons H_2O + CO$$

(3) $$3H_2 + N_2 \rightleftharpoons 2NH_3$$

The equilibrium constants for these reactions at $T = 298^\circ K$ are

$$K_1 = \frac{p_{CH_4} p_{H_2O}^2}{p_{CO_2} p_{H_2}^4} = 6 \times 10^{19}$$

$$K_2 = \frac{p_{H_2O} p_{CO}}{p_{H_2} p_{CO_2}} = 9.7 \times 10^4$$

$$K_3 = \frac{p_{NH_3}^2}{p_{H_2}^3 p_{N_2}} = 6.17 \times 10^6$$

We have three equations and seven unknown concentrations. The other four equations necessary to completely specify the system are mass-balance requirements.   They are

(4)′   Total N $= n_N = 2n_{N_2} + n_{NH_3}$

(5)′   Total H $= n_H = 4n_{CH_4} + 2n_{H_2} + 2n_{H_2O} + 3n_{NH_3}$

(6)′   Total O $= n_O = n_{H_2O} + n_{CO} + 2n_{CO_2}$

(7)′   Total C $= n_C = n_{CH_4} + n_{CO} + n_{CO_2}$

Since we are assuming all the gases to be ideal, we may express the mass balance in terms of partial pressures:

(4)   N: $2p_{N_2} + p_{NH_3} = \text{const}$

(5)   H: $4p_{CH_4} + 2p_{H_2} + 2p_{H_2O} + 3p_{NH_3} = \text{const}$

(6)   O: $p_{H_2O} + p_{CO} + 2p_{CO_2} = \text{const}$

(7)   C: $p_{CH_4} + p_{CO} + p_{CO_2} = \text{const}$

In order to proceed we must specify the temperature, total pressure, and starting mixture of gases.   Let the total pressure $= 1$ atm.   The starting mixture will consist of CH_4, H_2O, CO, and NH_3, with

$$p_{CO} = 0.1 \text{ atm}$$

$$p_{CH_4} = 0.1 \text{ atm}$$

$$p_{NH_3} = 0.4 \text{ atm}$$

$$p_{H_2O} = 0.4 \text{ atm}$$

We can now calculate the concentration of each gas in the final mixture. The mass-balance equations become

$$2p_{N_2} + p_{NH_3} = 0.4 \text{ atm}$$

$$4p_{CH_4} + 2p_{H_2} + 2p_{H_2O} + 3p_{NH_3} = 2.4 \text{ atm}$$

$$p_{H_2O} + p_{CO} + 2p_{CO_2} = 0.5 \text{ atm}$$

$$p_{CH_4} + p_{CO} + p_{CO_2} = 0.2 \text{ atm}$$

Now we have seven simultaneous equations in seven unknowns.   There are rather intricate methods for solving a system of seven equations, but it is at best a grizzly task.   A great deal of effort can be avoided by applying a bit of common sense and intuition.   Since K_1 is so large, let us assume that essentially all the C is present as CH_4 and essentially all the O_2 as H_2O. Thus

$$p_{CH_4} \cong 0.2 \text{ atm}$$

$$p_{H_2O} \cong 0.5 \text{ atm}$$

Assuming p_H to be negligible,

$$p_{NH_3} = \tfrac{1}{3}(2.4 \text{ atm} - 2p_{H_2O} - 4p_{CH_4})$$

$$= \tfrac{1}{3}(2.4 - 1.0 - 0.8)$$

$$= 0.2 \text{ atm}$$

Then p_{N_2} follows immediately:

$$p_{N_2} = \tfrac{1}{2}(0.4 \text{ atm} - p_{NH_3})$$

$$= 0.1 \text{ atm}$$

We may now check our assumptions by calculating the amounts of H_2, CO, and CO_2 from the equilibrium constants:

$$\frac{p_{NH_3}{}^2}{p_{H_2}{}^3 p_{N_2}} = \frac{(0.2)^2}{p_{H_2}{}^3 (0.1)} = 6.17 \times 10^6$$

$$p_{H_2}{}^3 = 6.48 \times 10^{-8}$$

$$p_{H_2} = 4.0 \times 10^{-3}$$

The CO_2 content is calculated from K_1:

$$\frac{p_{CH_4} p_{H_2O}{}^2}{p_{CO_2} p_{H_2}{}^4} = \frac{(0.2)(0.5)^2}{p_{CO_2}(4.0 \times 10^{-3})^4} = 6 \times 10^{19}$$

$$p_{CO_2} = 3.26 \times 10^{-12}$$

The final CO concentration, p_{CO}, may be calculated from K_2:

$$\frac{p_{H_2O} p_{CO}}{p_{H_2} p_{CO_2}} = \frac{(0.5) p_{CO}}{(4.0 \times 10^{-3})(3.26 \times 10^{-12})} = 9.7 \times 10^4$$

$$p_{CO} = 2.53 \times 10^{-9}$$

Our original guesses concerning the major components are entirely justified. The final mixture will contain

$$p_{CH_4} = 0.2 \text{ atm}$$
$$p_{H_2O} = 0.5 \text{ atm}$$
$$p_{NH_3} = 0.2 \text{ atm}$$
$$p_{N_2} = 0.1 \text{ atm}$$
$$p_{H_2} = 4.0 \times 10^{-3} \text{ atm}$$
$$p_{CO_2} = 3.26 \times 10^{-12} \text{ atm}$$
$$p_{CO} = 2.53 \times 10^{-9} \text{ atm}$$

NONIDEAL GASES

The behavior of real gases can be approximated for low pressures by the ideal gas law, but as the pressure increases real gases become more and more nonideal. The accuracy of equilibrium calculations involving water vapor at 50 atm is seriously impaired if ideal gas behavior is assumed. Problems encountered in igneous and metamorphic petrology require accurate calculations of equilibria involving gases such as CO_2 and H_2O at high pressures. To treat such systems we must use the concept of *fugacity*.

The fugacity, or escaping tendency, of a gas is defined by the equation

$$\mu = \mu^\circ + RT \ln f$$

where μ is the chemical potential of the pure gas at temperature T and pressure P, μ° is the chemical potential of the pure gas at temperature T when $f = 1$, and f is the fugacity.

To relate fugacity to pressure we require that

$$\frac{f}{P} \to 1 \quad \text{as} \quad P \to 0$$

so that the fugacity equals the pressure under conditions where the gas behaves ideally.

For this relationship to be useful, we must be able to calculate the fugacity from either experimental data or an equation of state. This may be done as follows: At constant temperature

$$d\mu = V \, dP$$

where V = volume per mole and, since μ° is a function of temperature only,

$$d\mu = RT \, d \ln f$$

Thus

$$RT \, d \ln f = V \, dP$$

Let us subtract $RT \, d \ln P$ from both sides of this equation, which gives

$$RT \, d \ln f - RT \, d \ln P = V \, dP - RT \, d \ln P$$

$$RT \, d \ln \frac{f}{P} = \left(V - \frac{RT}{P} \right) dP$$

The term $V - RT/P$ is the real volume of the gas minus the volume which the gas would occupy if it were ideal. Integrating this equation between states P_1 and P_2 at constant temperature gives

$$RT \ln \left(\frac{f_2/P_2}{f_1/P_1} \right) = \int_{P_1}^{P_2} \left(V - \frac{RT}{P} \right) dP$$

Both H_2O and CO_2 are sufficiently ideal at 1 atm so that we may set $P_1 = 1$ atm and $f_1/P_1 = 1$. Thus

$$RT \ln f = RT \ln P + \int_1^P \left(V - \frac{RT}{P} \right) dP$$

or

$$\ln f = \ln P + \frac{1}{RT} \int_1^P (V_{\text{real}} - V_{\text{ideal}}) \, dP$$

We could now calculate fugacity from measurements of the pressure-volume-temperature (P-V-T) behavior of the gas.

We may also calculate fugacity from the equation of state for the gas. For pressures up to 200 atm for water vapor at high temperature we may use the van der Waals equation of state. It is convenient when using the van der Waals equation to change the integration variable from pressure to volume:

$$P = \frac{RT}{V - b} - \frac{a}{V^2}$$

$$\frac{dP}{dV} = -\frac{RT}{(V - b)^2} + \frac{2a}{V^3}$$

Let $V_1 =$ volume of gas when $P = 1$.   Then

$$\ln f = \frac{1}{RT} \int_{V_1}^{V_P} V \left[-\frac{RT}{(V - b)^2} + \frac{2a}{V^3} \right] dV$$

and

$$\ln f = \frac{1}{RT} \left[-\frac{2a}{V} + \frac{RTV}{V - b} - RT \ln (V - b) \right]_{V_1}^{V_P}$$

The symbols outside the brackets mean that the expression is evaluated at V_P and at V_1.   The desired result ($\ln f$) is the value at V_P minus the value at V_1.

We have used this expression to calculate the fugacity of water vapor at 400°C (673°K) at pressures up to 200 atm.   The results are given in Table 6.1.   (We have tabulated f/p, which is called the activity

TABLE 6.1

Activity Coefficient for Water Vapor at 400°C

Volume, liters	Pressure, atm	f/p (calculated)	f/p† (measured)
20	2.75	0.9926	
5	10.88	0.983	
2	26.65	0.964	
1	51.43	0.933	
0.8	63.13	0.918	
0.6	81.64	0.897	
~0.5	98.7	0.876	0.874
0.4	115.0	0.856	
0.3	143.7	0.821	
0.2	188.0	0.763	
0.18	199.3	0.751	0.737

† Data from W. T. Holser, *J. Phys. Chem.*, **58**: 316 (1954).

coefficient.)   Experimental values at 100 and 200 bars taken from the data of Holser are shown.   Since 1 bar = 0.9869 atm, these points correspond to 98.7 and 197.9 atm.   At 99 atm the agreement is almost perfect; at 198 atm an error of only $1\frac{1}{2}$ percent would result from using the calculated values.   Above 200 atm the error would continue to increase, and so at very high pressures experimental values must be used. It should be noted that the deviation from ideality for water is already 7 percent at 50 atm.   This could produce large errors in calculating the equilibrium composition of a mixture containing H_2O at 50 atm.

MIXTURES OF NONIDEAL GASES

The fugacity of a nonideal gas in a mixture can be defined in a manner
analogous to that for pure nonideal gas. The chemical potential of the
gas in a mixture is defined as

$$\mu_i = \mu_i^\circ + RT \ln f_i$$

The quantity μ_i° is the chemical potential of pure gas, i, at unit fugacity.
Since

$$d\mu_i = \bar{V}_i \, dP$$

where $\bar{V}_i$ is the partial molal volume of i and

$$d\mu_i = RT \, d \ln f_i$$

we have

$$RT \, d \ln f_i = \bar{V}_i \, dP$$

Subtracting $RT \, d \ln p_i$ from both sides,

$$RT \, d \ln f_i - RT \, d \ln p_i = \bar{V}_i \, dP - RT \, d \ln p_i$$

But $p_i = X_i P$, and since the mole fraction, X_i, of i is constant

$$d \ln p_i = d \ln P$$

Thus

$$RT \, d \ln \frac{f_i}{p_i} = \left(\bar{V}_i - \frac{RT}{P} \right) dP$$

and

$$\ln \frac{f_i}{p_i} = \frac{1}{RT} \int_1^P \left(\bar{V}_i - \frac{RT}{P} \right) dP$$

Under certain conditions the fugacities of real gases follow a law
similar to the law of partial pressures. This law was first postulated by
G. N. Lewis and M. Randall and is known as Lewis and Randall's rule.
It states that the fugacity of i in a mixture equals the fugacity of pure i
at the same temperature and *total* pressure multiplied by the mole fraction of i in the mixture. Thus

$$f_i \text{ (at } T, X_i P) = X_i f_i^\circ \text{ (at } T, P)$$

The physical requirement for gases to follow this rule is that the chief
source of nonideality must stem from interaction between like molecules

and not from interactions between the different components of the mixture. A mixture of CO_2 and H_2O might be expected to follow Lewis and Randall's rule if the interactions between H_2O and H_2O molecules and CO_2 and CO_2 molecules are the main cause of nonideality and the CO_2-H_2O interactions produce only a small additional nonideality. If the CO_2-H_2O interactions produce a large amount of additional nonideality, this rule cannot be used.

It can be shown that gaseous solutions for which Lewis and Randall's fugacity rule is valid have two additional properties.

1. For each component the partial molal volume is equal to the molar volume.
2. There is no heat of mixing.

Since fugacity data for mixtures of gases are rare, the Lewis and Randall rule is used for some systems where it might not be a good approximation to reality. The degree to which the volume and heat-of-mixing conditions are violated can give a qualitative estimate of the reliability of this rule for a given gaseous mixture.

THE TRUE EQUILIBRIUM CONSTANT

The equilibrium constant for reactions involving real gases may be derived from the expression for chemical potential:

$$\mu_i = \mu_i^\circ + RT \ln f_i$$

A line of argument identical to that used for ideal gases leads to the result

$$- RT \ln \Pi f_i^{\nu_i} = \Sigma \nu_i \mu_i^\circ$$

We define the true equilibrium constant K_f to be

$$K_f \equiv \Pi f_i^{\nu_i}$$

Thus, since $\Sigma \nu_i \mu_i^\circ = \Delta G_T^\circ$,

$$\ln K_f = - \frac{\Delta G_T^\circ}{RT}$$

The change in K_f with temperature is given by

$$\frac{d \ln K_f}{dT} = \frac{\Delta H^\circ}{RT^2}$$

The equilibrium constant expressed in terms of fugacity is completely independent of pressure.

The relationship between K_f and K_p can best be illustrated by an example. Consider the reaction

$$2CO + O_2 \rightarrow 2CO_2$$

The equilibrium constant in terms of partial pressures is

$$K_p = \frac{p_{CO_2}^2}{p_{CO}^2 p_{O_2}}$$

and in terms of fugacity,

$$K_f = \frac{f_{CO_2}^2}{f_{CO}^2 f_{O_2}}$$

The ratio f/p is the activity coefficient, Γ, for the gas.   Thus

$$K_f = \frac{(f/p)_{CO_2}^2}{(f/p)_{CO}^2 (f/p)_{O_2}} \frac{p_{CO_2}^2}{p_{CO}^2 p_{O_2}}$$

$$K_f = \frac{\Gamma_{CO2}^2}{\Gamma_{CO}^2 \Gamma_{O_2}} K_p$$

Since the activity coefficients vary with pressure (see Table 6.1, for example), K_p must also vary with pressure for real gases.

PROBLEMS

6.1 Assuming ideal gases, calculate the production of NH_3 from equal amounts (number of moles) of N_2 and H_2 at 1 atm total pressure and 10 atm total pressure.   Explain your result.

*6.2 The abundance of elements in the solar system is given by H. E. Suess and H. C. Urey [*Rev. Mod. Phys.*, **28**: 53 (1956)] relative to 10^4 atoms of Si as

$$H = 4.0 \times 10^8 \qquad O = 2.15 \times 10^5$$
$$He = 3.1 \times 10^7 \qquad Ne = 8.6 \times 10^4$$
$$C = 3.5 \times 10^4 \qquad S = 3.75 \times 10^3$$
$$N = 6.6 \times 10^4$$

Calculate the abundance of all simple chemical compounds such as CO, CO_2, O_2, CH_4, H_2O, H_2, NH_3, N_2, H_2S, and S in a gas of this chemical composition at $T = 1500°C$, $P = 10^{-3}$ atm and at $T = 1200°C$, $P = 10^{-3}$ atm. Free energy data at 1 atm, 298°K may be found in R. M. Garrels and C. L. Christ, "Solutions, Minerals, and Equilibria" (Harper & Row, Publishers, Incorporated, New York, 1965). Heat-capacity data are in K. K. Kelley (*U.S. Bur. Mines Bull.* 584, 1960). Assume all gases are ideal and that only those compounds listed can form.

*6.3 Volcanic gases from Surtsey, Iceland, have been studied by Sigvaldason and Elisson. One set of measurements showed the following composition in mole percent at $T = 1400°K$:

$$H_2O = 88.16\% \qquad H_2 = 4.74\%$$
$$HCl = 0.40 \qquad CO = 0.38$$
$$SO_2 = 3.28 \qquad N_2 + Ar = 0.07$$
$$CO_2 = 4.97 \qquad O_2 = 0$$

What is the probable origin of the N_2 content? Calculate the value of the equilibrium constant for the reaction

$$CO_2 + H_2 = CO + H_2O$$

at 1400°K given that $K_p = 1.4$ at 1200°K. Is the observed gas composition in equilibrium at 1400°K? Would your answer be different if the equilibration pressure were 100 atm?

6.4 Using the van der Waals equation, calculate the pressure, fugacity, and f/p for CO_2 gas at $T = 400°C$ and $V = 20$ liters, $V = 2$ liters, $V = 1$ liter, $V = 0.5$ liter, $V = 0.3$ liter, and $V = 0.2$ liter.

6.5 Plot the data in Table 6.1 as fugacity versus pressure for H_2O; do the same for the results of Prob. 6.4. Using the Lewis and Randall rule, calculate the fugacity of H_2O and CO_2 in a mixture where $x_{H_2O} = 0.9$ and $x_{CO_2} = 0.1$ at 400°C and pressures of 50, 100, 150, and 200 atm.

*6.6 In automobile exhaust the molar ratio of CO/CO_2 is 0.1. In the atmosphere, as of the year 1900, the CO_2 pressure was 3×10^{-4} atm, the H_2O pressure was 1×10^{-2}, and the H_2 pressure was 0.5×10^{-6} atm. Under these conditions, if mean air temperature is taken to be 15°C, what is the equilibrium CO/CO_2 ratio?

SUPPLEMENTARY READING

Eck, R. V., E. R. Lippincott, M. O. Dayhoff, and Y. T. Pratt: Thermodynamic Equilibrium and the Inorganic Origin of Organic Compounds, *Science,* **153:** 628 (1966).

Hayatsu, R., M. H. Studier, A. Oda, K. Fuse, and E. Anders: Origin of Organic Matter in Early Solar System, II, Nitrogen Compounds, *Geochim. Cosmochim. Acta,* **32:** 175 (1968).

Leovy, C. B.: Atmospheric Ozone: An Analytic Model for Photochemistry in the Presence of Water Vapor, *J. Geophys. Res.,* **74:** 417 (1969).

McClaine, L. A., R. V. Allen, R. K. McConnell, Jr., and N. F. Surprenant: Volcanic Smoke Clouds, *J. Geophys. Res.,* **73:** 5235 (1968).

Sigvaldason, G. E., and G. Elisson: Collection and Analysis of Volcanic Gases of Surtsey, Iceland, *Geochim. Cosmochim. Acta,* **32:** 797 (1968).

chapter seven

Distribution of Trace Isotopes between Coexisting Phases

Several light elements such as O_2, S, C, N_2, H_2, and Si have stable trace isotopes which show variable abundances in natural materials. This variability is caused by isotope fractionation during chemical reactions. Since all the isotopes of a given element have identical electronic structures, the bonds they form in any given compound are identical. However, the energy levels for vibration depend on mass as well as bond strength, so that an isotopic substitution leads to a change in vibration frequency. The mass difference also influences translational and rotational motions. Consequently, a molecule containing O^{18} differs in its thermal properties, such as internal energy, heat capacity, and entropy, from one containing O^{16}. These differences lead to small equilibrium fractionations of isotopes between coexisting phases. Although the energy wells for two isotopes of the same element are identical, the zero-point energy is greater for a molecule made up of light atoms, as shown in Fig. 7.1. This means that bonds formed by the light isotope are more readily broken than bonds involving the heavy isotope. Thus, during a chemical reaction, molecules bearing the light isotope will, in general, react slightly more readily than those with the heavy isotope.

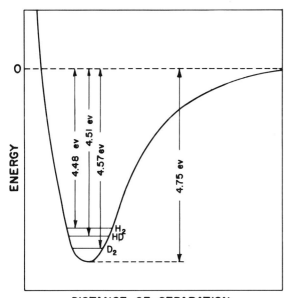

DISTANCE OF SEPARATION

Fig. 7.1. Schematic diagram showing the relationship between the zero-point energy and molecular mass for hydrogen (H_2), deuterium (D_2), and HD. The fundamental vibration frequencies are H_2: 4405 cm^{-1}, HD: 3817 cm^{-1}, D_2: 3119 cm^{-1}. The zero-point energy of H_2 is greater than that for HD which is greater than that for D_2.

In addition to equilibrium fractionations, other kinetic separations may also occur. For example, in a gas the molecules containing the light isotope move more rapidly than those with the heavy isotope. This velocity difference leads to isotope separation during diffusion.

Variations in the isotopic composition of the elements in natural materials result from both equilibrium and kinetic fractionations. In this chapter we will discuss the calculation of equilibrium-fractionation factors. Kinetic fractionation, which can be derived only from measurements on natural systems, will be discussed only briefly.

EQUILIBRIUM FRACTIONATION FOR REACTIONS INVOLVING DIATOMIC GASES

In order to understand the nature of equilibrium isotopic separations, let us consider an example. Two moles of carbon monoxide gas, CO, are equilibrated with 1 mole of oxygen gas, O_2. Assume that no chemical reaction takes place between the two molecular species and that the trace amount of O^{18} present (1 atom in 500) achieves an equilibrium distribution between the two molecular types. A separation constant, α, can be defined such that

$$\alpha = \frac{(O^{18}/O^{16})_{CO}}{(O^{18}/O^{16})_{O_2}}$$

In terms of the five isotopic species present in the system ($O^{16}O^{16}$, $O^{16}O^{18}$, $O^{18}O^{18}$, CO^{16}, and CO^{18}) the separation constant becomes

$$\alpha = \frac{[CO^{18}]/[CO^{16}]}{(2[O^{18}O^{18}] + [O^{16}O^{18}])/(2[O^{16}O^{16}] + [O^{16}O^{18}])}$$

The concentrations of these five species are related by the equilibrium constants for the following isotopic reactions:

$$CO^{16} + O^{18}O^{16} = CO^{18} + O^{16}O^{16}$$

and

$$O^{18}O^{18} + O^{16}O^{16} = 2O^{16}O^{18}$$

The equilibrium constants are as follows:

$$K_1 = \frac{[O^{16}O^{16}][CO^{18}]}{[O^{18}O^{16}][CO^{16}]}$$

and

$$K_2 = \frac{[O^{16}O^{18}]^2}{[O^{18}O^{18}][O^{16}O^{16}]}$$

Concentrations rather than fugacities are used since the constants relating fugacity to concentration in the numerators of the expressions would be exactly canceled by those in the denominators. (As the two species are chemically identical, $\Gamma_{CO^{16}} = \Gamma_{CO^{18}}$ and so forth.)

From these relationships we can show that

$$\frac{[CO^{18}]}{[CO^{16}]} = K_1 \frac{[O^{18}O^{16}]}{[O^{16}O^{16}]}$$

and

$$[O^{18}O^{18}] = \frac{[O^{16}O^{18}]^2}{K_2[O^{16}O^{16}]}$$

Substituting in the equation for α,

$$\alpha = \frac{K_1([O^{16}O^{18}]/[O^{16}O^{16}])}{\dfrac{(2/K_2)([O^{16}O^{18}]^2/[O^{16}O^{16}]) + [O^{16}O^{18}]}{2[O^{16}O^{16}] + [O^{16}O^{18}]}}$$

and, rearranging terms,

$$\alpha = 2K_1 \frac{2[O^{16}O^{16}] + [O^{16}O^{18}]}{2[O^{16}O^{16}] + (4/K_2)([O^{16}O^{18}])}$$

The value of K_2 will turn out to be almost exactly 4 at all temperatures (to be proved below); hence the separation factor is equal to twice the equilibrium constant for the isotopic reaction between CO and O_2 (that is, $\alpha = 2K_1$).

The problem is then to establish this equilibrium constant. As for all reactions at equilibrium, the free energy of the reactants must equal the free energy of the products:

$$\Delta G = 0 = G_{O^{16}O^{16}} + G_{CO^{18}} - G_{O^{18}O^{16}} - G_{CO^{16}}$$

The free energy for each gas can be written

$$G(T,P) = G^\circ(T) + RT \ln p$$

Using this relationship,

$$\Delta G^\circ = -RT \ln \frac{p_{O^{16}O^{16}} p_{CO^{18}}}{p_{O^{18}O^{16}} p_{CO^{16}}}$$

Since the isotopic reaction does not alter the number of molecules present, concentrations can be directly substituted for partial pressures:

$$\Delta G^\circ = -RT \ln \frac{[O^{16}O^{16}][CO^{18}]}{[O^{18}O^{16}][CO^{16}]} = -RT \ln K_1$$

Solving for K_1,

$$K_1 = \exp\left(-\frac{\Delta G^\circ}{RT}\right)$$

The next step is to determine the free energies for the isotopic species. As always,

$$\Delta G = \Delta H - T\,\Delta S$$

Since there is no volume change in connection with isotopic reactions, the enthalpy change, ΔH, must be the same as the internal-energy change, ΔE. Thus

$$\Delta G = \Delta E - T\,\Delta S$$

In Chap. 4 we showed from statistical mechanics that

$$S = \frac{E}{T} + R \ln q$$

Hence

$$T \, \Delta S = \Delta E + RT \ln \frac{q_{O^{16}O^{16}} q_{CO^{18}}}{q_{O^{18}O^{16}} q_{CO^{16}}}$$

Finally

$$\Delta G^\circ = -RT \ln \left(\frac{q^\circ_{O^{16}O^{16}} q^\circ_{CO^{18}}}{q^\circ_{O^{18}O^{16}} q^\circ_{CO^{16}}} \right)_T$$

The equilibrium constant becomes

$$K_1 = \left(\frac{q_{O^{16}O^{16}} q_{CO^{18}}}{q_{O^{18}O^{16}} q_{CO^{16}}} \right)_T$$

The subscript T indicates that the equilibrium constant holds only for the specific temperature used in the calculation. This is because the partition functions vary with temperature. Since

$$q = q_{\text{trans}} \frac{e}{N} q_{\text{rot}} \, q_{\text{vib}}$$

we can calculate separately the fractionations due to translation, rotation, and vibration.

TRANSLATIONAL COMPONENT

As shown in Chap. 4, the partition function for translation is given by

$$q_{\text{trans}} = \left(\frac{2\pi m k T}{h^2} \right)^{\frac{3}{2}} V$$

Thus, since V is the same for each species,

$$\left(\frac{q_{CO^{18}}}{q_{CO^{16}}} \right)_{\text{trans}} = \left(\frac{m_{CO^{18}}}{m_{CO^{16}}} \right)^{\frac{3}{2}} = \left(\frac{30}{28} \right)^{\frac{3}{2}}$$

$$\left(\frac{q_{O^{16}O^{16}}}{q_{O^{16}O^{18}}} \right)_{\text{trans}} = \left(\frac{m_{O^{16}O^{16}}}{m_{O^{16}O^{18}}} \right)^{\frac{3}{2}} = \left(\frac{32}{34} \right)^{\frac{3}{2}}$$

Finally

$$K_{\text{trans}} = \left(\tfrac{30}{28} \tfrac{32}{34} \right)^{\frac{3}{2}} = \left(\tfrac{120}{119} \right)^{\frac{3}{2}} = 1.0126$$

Were the difference in translational energy modes the only source of fractionation, the trace isotope O^{18} would be 13 per mil more abundant in the CO (the light molecule) than in the O_2 (the heavier molecule). This fractionation would not change with temperature.

ROTATIONAL COMPONENT

For rotation the partition function is given by

$$q_{\mathrm{rot}} = \frac{8\pi^2 \mu d^2 kT}{\sigma h^2}$$

Thus, since the $O^{16}O^{18}$ and $O^{16}O^{16}$ bond lengths are identical,

$$\left(\frac{q_{O^{16}O^{16}}}{q_{O^{16}O^{18}}}\right)_{\mathrm{rot}} = \frac{1}{2}\frac{\mu_{O^{16}O^{16}}}{\mu_{O^{16}O^{18}}}$$

$$= \frac{1}{2}\frac{(16 \times 16)/(16 + 16)}{(16 \times 18)/(16 + 18)} = \frac{1}{2}\frac{17}{18}$$

The factor of $\frac{1}{2}$ is necessary because symmetric molecules have only half the rotational modes of asymmetric molecules.   This factor of $\frac{1}{2}$ cancels the factor of 2 in the relationship $\alpha = 2K_1$.

For CO,

$$\left(\frac{q_{CO^{18}}}{q_{CO^{16}}}\right)_{\mathrm{rot}} = \frac{\mu_{CO^{18}}}{\mu_{CO^{16}}} = \frac{(12 \times 18)/(12 + 18)}{(12 \times 16)/(12 + 16)} = \frac{21}{20}$$

Thus

$$K_{\mathrm{rot}} = \tfrac{1}{2}\tfrac{21}{20}\tfrac{17}{18} = \tfrac{1}{2}\tfrac{119}{120} = \tfrac{1}{2}(0.9916)$$

Were the difference in the rotational energy modes the only source of fractionation, O^{18} would be 8 per mil more abundant in the O_2 than in the CO.   Again, the fractionation does not depend on temperature.

Taking the rotational and translational effects together, the two effects oppose each other, giving

$$K_{\mathrm{trans}}\,K_{\mathrm{rot}} = \tfrac{1}{2}\sqrt{\tfrac{120}{119}} = \tfrac{1}{2}(1.0042)$$

VIBRATIONAL COMPONENT

The vibrational partition function is given by

$$q_{\mathrm{vib}} = \frac{\exp(-h\nu/2kT)}{1 - \exp(-h\nu/kT)}$$

At temperatures well below $h\nu/k$, the partition function can be approximated by

$$q_{\text{vib}} = \exp\left(-\frac{h\nu}{2kT}\right)$$

Thus, for CO,

$$\left(\frac{q_{CO^{18}}}{q_{CO^{16}}}\right)_{\text{vib}} = \exp\left[-\frac{h}{2kT}\left(\nu_{CO^{18}} - \nu_{CO^{16}}\right)\right]$$

and for O_2

$$\left(\frac{q_{O^{16}O^{16}}}{q_{O^{16}O^{18}}}\right)_{\text{vib}} = \exp\left[-\frac{h}{2kT}\left(\nu_{O^{16}O^{16}} - \nu_{O^{16}O^{18}}\right)\right]$$

Finally

$$K_{\text{vib}} = \exp\left\{+\frac{h}{2kT}\left[\left(\nu_{CO^{16}} - \nu_{CO^{18}}\right) - \left(\nu_{O^{16}O^{16}} - \nu_{O^{16}O^{18}}\right)\right]\right\}$$

Since the fractionations are in general only a few percent, we may approximate the vibrational equilibrium constant by

$$K_{\text{vib}} \cong 1 + \frac{h}{2kT}\left[\left(\nu_{CO^{16}} - \nu_{CO^{18}}\right) - \left(\nu_{O^{16}O^{16}} - \nu_{O^{16}O^{18}}\right)\right]$$

For an ideal harmonic oscillator

$$\nu = \frac{1}{2\pi}\sqrt{\frac{k}{\mu}}$$

Since the force constants, k, do not differ for the isotopically substituted molecules,

$$\nu_{CO^{18}} = \nu_{CO^{16}}\sqrt{\frac{\mu_{CO^{16}}}{\mu_{CO^{18}}}} = \sqrt{\frac{20}{21}}\,\nu_{CO^{16}}$$

and

$$\nu_{O^{16}O^{18}} = \nu_{O^{16}O^{16}}\sqrt{\frac{\mu_{O^{16}O^{16}}}{\mu_{O^{16}O^{18}}}} = \sqrt{\frac{17}{18}}\,\nu_{O^{16}O^{16}}$$

Thus

$$K_{\text{vib}} \cong 1 + \frac{h}{2kT}\left[\nu_{CO^{16}}\left(1 - \sqrt{\frac{20}{21}}\right) - \nu_{O^{16}O^{16}}\left(1 - \sqrt{\frac{17}{18}}\right)\right]$$

$$\cong 1 + \frac{h}{2kT}\left(0.0244\nu_{CO^{16}} - 0.0282\nu_{O^{16}O^{16}}\right)$$

Laboratory measurements give

$$\nu_{CO^{16}} = 6.50 \times 10^{13} \text{ sec}^{-1}$$
$$\nu_{O^{16}O^{16}} = 4.74 \times 10^{13} \text{ sec}^{-1}$$

Since $h/2k = 2.4 \times 10^{-11}$ deg-sec, we have

$$K_{\text{vib}} = 1 + \frac{2.4 \times 10^{-11}}{T} (15.86 - 13.37)$$

$$= 1 + \frac{5.976}{T}$$

At room temperature the separation factor for vibration becomes

$$K_{\text{vib}} = 1 + \frac{5.976}{300} = 1.01992$$

Were the difference in vibrational energy modes the only source of fractionation, the trace isotope O^{18} would, at room temperature, be 20 per mil enriched in the CO, the molecule with the higher vibrational frequency. This fractionation would increase by roughly 0.06 per mil for each degree the temperature was decreased and would fall by the same amount for each degree the temperature increased.

The overall separation factor becomes, at room temperature,

$$\alpha = 2K_1$$
$$= 2K_{\text{trans}}K_{\text{rot}}K_{\text{vib}}$$
$$= (2)(\tfrac{1}{2})(1.0042)(1.0199) = 1.024$$

At equilibrium, there would be a 24 per mil enrichment of O^{18} in the CO relative to the O_2.

Let us now check our assumption that $K_2 = 4$. The translational contribution will be

$$K_{\text{trans}} = \left[\frac{(m_{O^{16}O^{18}})^2}{m_{O^{16}O^{16}}m_{O^{18}O^{18}}} \right]^{\frac{3}{2}} = \left[\frac{(34)(34)}{(32)(36)} \right]^{\frac{3}{2}} = \left(\frac{289}{288} \right)^{\frac{3}{2}}$$

The rotational contribution will be

$$K_{\text{rot}} = \frac{(\mu_{O^{16}O^{18}})^2}{(\tfrac{1}{2}\mu_{O^{16}O^{16}})(\tfrac{1}{2}\mu_{O^{18}O^{18}})}$$

$$= 4\frac{(32)(36)}{(34)(34)} = 4\frac{288}{289}$$

The product of these two components will be

$$K_{\text{trans}}K_{\text{rot}} = 4\sqrt{\tfrac{289}{288}} = 4(1.0017)$$

Finally the vibrational contribution will be

$$K_{\text{vib}} = \exp\left[-\frac{h}{2kT}\left(2\nu_{O^{16}O^{18}} - \nu_{O^{16}O^{16}} - \nu_{O^{18}O^{18}}\right)\right]$$

Since

$$\nu_{O^{18}O^{18}} = \nu_{O^{16}O^{16}}\sqrt{\frac{\mu_{O^{16}O^{16}}}{\mu_{O^{18}O^{18}}}} = \sqrt{\frac{8}{9}}\,\nu_{O^{16}O^{16}}$$

and

$$\nu_{O^{16}O^{18}} = \nu_{O^{16}O^{16}}\sqrt{\frac{\mu_{O^{16}O^{16}}}{\mu_{O^{16}O^{18}}}} = \sqrt{\frac{17}{18}}\,\nu_{O^{16}O^{16}}$$

we have

$$K_{\text{vib}} = \exp\left[-\frac{h}{2kT}\,\nu_{O^{16}O^{16}}\left(2\sqrt{\tfrac{17}{18}} - 1 - \sqrt{\tfrac{8}{9}}\right)\right]$$

$$= \exp\left(-\frac{0.0008h\nu_{O^{16}O^{16}}}{2kT}\right)$$

$$= 1 - \frac{0.0008h\nu_{O^{16}O^{16}}}{2kT}$$

$$= 1 - \frac{0.846}{T}$$

If $T = 300°\text{K}$,

$$K_{\text{vib}} = 0.9972$$

The final value of K at $300°\text{K}$ is

$$K = K_{\text{trans}}K_{\text{rot}}K_{\text{vib}}$$
$$= 4(1.0017)(0.9972)$$
$$= 3.996$$

Returning to the relationship between α and K, we have

$$\alpha = 2K_1\,\frac{2[O^{16}O^{16}] + [O^{16}O^{18}]}{2[O^{16}O^{16}] + (4.00/3.996)[O^{16}O^{18}]}$$

Since in natural systems there is roughly one $O^{16}O^{18}$ molecule per 500

$O^{16}O^{16}$ molecules

$$\alpha \cong 2K_1 \frac{1,001,000}{1,001,001}$$

The correction is only 1 ppm and hence can be neglected.

As mentioned above, the translational and rotational components of the fractionation do not vary with temperature. At low temperatures the vibrational component varies inversely with the absolute temperature. As the temperature increases, the $1 - \exp(-h\nu/kT)$ term in the denominator of the partition function becomes important. The equilibrium constant, K, is then given by

$$K_{vib} = \exp\left\{-\frac{h\,\Delta\nu}{2kT}\frac{[1 - \exp(-h\nu_{O^{16}O^{18}}/kT)][1 - \exp(-h\nu_{CO^{16}}/kT)]}{[1 - \exp(-h\nu_{O^{16}O^{16}}/kT)][1 - \exp(-h\nu_{CO^{18}}/kT)]}\right\}$$

The temperature dependence of the equilibrium constant is shown in Fig. 7.2. The contribution of the $1 - \exp(-h\nu/kT)$ term to the fractionation is zero at absolute zero, rises to a broad maximum centered at about 2500°K, and then gradually decreases. At infinite temperature the vibrational fractionation just balances the translation and rotation term.

The question arises as to why O^{18} is concentrated in the CO as a result of translational and vibrational motion and in the O_2 as a result of rotational motion. Although the answer must certainly be that these concentrations lead to a higher entropy for the universe than would be the case if the O^{18} were uniformly distributed, it is not so obvious why this is true. We made the calculation as follows: The free energy difference between the reactants $(O^{18}O^{16} + CO^{16})$ and the products $(O^{16}O^{16} + CO^{18})$ was determined for the case when all four gases were present in equal amounts. Under these conditions we found that the number of translational states available to the products was proportional to the product of their masses to the $\frac{3}{2}$ power. The same was, of course, true for the reactants. As the product of the masses of the reactants [that is, $(28)(34) = 952$] was smaller than that for the products [that is, $(30)(32) = 960$], the entropy of the products was slightly $[(960/952)^{\frac{3}{2}}]$ greater than that for the reactants. Under these conditions $CO^{16} + O^{18}O^{16}$ would react to yield additional CO^{18} and $O^{16}O^{16}$. The reaction would cause the partial pressure of the reactants to fall slightly and that of the products to rise slightly. As the density of CO^{18} and $O^{16}O^{16}$ molecules increased, their entropy would decrease. Similarly, entropy of the reactants would rise. This change would continue until the original entropy difference between the products and reactants was eliminated.

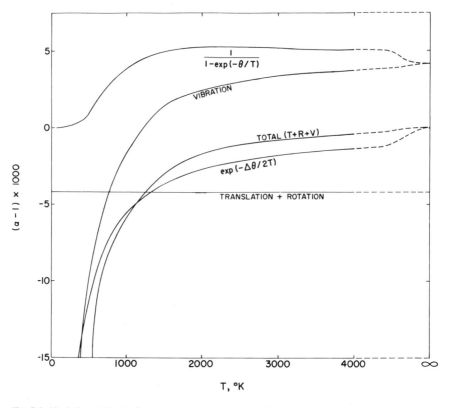

Fig. 7.2. Variation of isotopic fractionation of an ideal diatomic gas as a function of temperature. The contribution of the $1/[1 - \exp(-\theta/T)]$ part of the vibration term is zero at absolute zero, rises to a broad maximum, and then slowly decreases to approach a constant value at infinite temperature. At infinite temperature the vibrational fractionation exactly balances the translation and rotation components.

EQUILIBRIUM FRACTIONATION BETWEEN COEXISTING SILICATES

The isotope distribution between coexisting solid phases is controlled entirely by the vibrational frequencies in the solids. Translational and rotational motions do not, in general, take place in solids. Unfortunately the atoms in solids vibrate at many different frequencies. For most solids the exact spectrum of these frequencies is unknown. For this reason, separation factors cannot, in general, be calculated.

A rough approximation of separation factors can, however, be made. As pointed out in Chap. 3, the room-temperature heat capacity of any solid provides a measure of its "average" vibrational frequency. Quartz, whose heat capacity per atom is 3.54 cal/mole-atom, must be characterized by higher vibrational frequencies than hematite ($C_P/n = 4.96$ cal/mole-atom). It was further shown that the heat capacities of oxides were additive.   Quartz and silica glass have heat capacities differing by less than 1 percent.   The heat capacity of enstatite ($MgSiO_3$) is 19.62 compared with the sum 19.65(10.62 + 9.03) for quartz (SiO_2) and periclase (MgO).   Since little difference exists in the average vibration frequency, no isotope separation should be expected if pairs such as quartz–silica glass or enstatite–periclase + quartz were equilibrated.

The significance of this is as follows: No matter what complex oxides are involved, the calculation of separation factors can be made for the sum of the constituent oxides.   Thus, if the separation factor between sanidine ($KAlSi_3O_8$) and enstatite ($MgSiO_3$) were desired, the calculation would be made by assuming that a mixture of 1 part K_2O, 1 part Al_2O_3, and 6 parts SiO_2 was equilibrated with a mixture of 1 part MgO and 1 part SiO_2.   The SiO_2 portion of each batch would have the same isotopic composition.   If the fractionation factors for the pairs SiO_2-K_2O, SiO_2-MgO, and SiO_2-Al_2O_3 were known, the total separation factor could be quickly calculated.

Although this does not completely solve the problem, it greatly reduces the work necessary.   At each temperature of interest, instead of one measurement for each mineral, only one measurement need be made for each metal oxide.   Given the results for the oxides of Na, K, Ca, Mg, Fe^{++}, Fe^{3+}, Si, Al, and Ti, a large fraction of the rock-forming minerals would be covered (hydrous minerals and carbonates would be the main exceptions).

A rough estimate of the relative fractionations between these minerals can be obtained by assuming that each metal-oxygen pair acts as a diatomic molecule.   A typical reaction could then be written

$$AlO^{16} + SiO^{18} = AlO^{18} + SiO^{16}$$

The separation constant can be written

$$\alpha = \frac{(O^{18}/O^{16})_{Si-O}}{(O^{18}/O^{16})_{Al-O}}$$

$$= 1 - \frac{h}{2kT}\left[\nu_{SiO^{16}}\left(1 - \sqrt{\frac{\mu_{SiO^{16}}}{\mu_{SiO^{18}}}}\right) - \nu_{AlO^{16}}\left(1 - \sqrt{\frac{\mu_{AlO^{16}}}{\mu_{AlO^{18}}}}\right)\right]$$

where, for example,

$$\mu_{SiO^{16}} = \frac{m_{Si}m_{O^{16}}}{m_{Si} + m_{O^{16}}}$$

The vibration frequency is taken to be that which would yield the observed room-temperature heat capacity for the metal oxide were it characterized by a single frequency of oscillation.   By using Fig. 3.4 the corresponding value of θ_E can be found.   Then

$$\nu = \frac{k}{h}\,\theta_E$$

Values of ν computed in this way for each of the nine important oxides are given in Table 7.1, as are the values of $\mu_{O^{16}}$ and $1 - \sqrt{\mu_{16}/\mu_{18}}$.   The product of ν_{16} and $1 - \sqrt{\mu_{16}/\mu_{18}}$ should then be a measure of the tendency of each oxide to concentrate O^{18}.   The difference between this product for two different oxides should be a measure of the equilibrium fractionation.   By using quartz (i.e., that mineral with the highest vibrational frequency and hence the strongest tendency to enrich O^{18}) as a reference, the last column gives the calculated room-temperature equilibrium enrichment of O^{18} in quartz relative to each of the other oxides.

TABLE 7.1

Calculated Fractionations for Oxides Relative to Quartz

Oxide	C_P/n, cal/ mole-atom	θ_E, °K	ν, 10^{13} sec^{-1}	$\mu_{O^{16}}$	$1 - \sqrt{\dfrac{\mu_{16}}{\mu_{18}}}$	$\dfrac{h\nu}{2kT}\left(1 - \sqrt{\dfrac{\mu_{16}}{\mu_{18}}}\right)$† $\times 1000$	$\Delta^{\text{oxide}}_{\text{qtz}}$† $\times 1000$
SiO_2	3.54	770	1.64	10.2	0.036	47	0
Al_2O_3	3.78	715	1.52	9.9	0.035	42	-5
TiO_2	4.38	580	1.23	12.0	0.042	41	-6
MgO	4.52	555	1.18	9.6	0.034	32	-15
Fe_2O_3	4.97	450	0.96	12.7	0.043	33	-14
CaO	5.12	420	0.89	11.4	0.040	29	-18
FeO	5.25	375	0.80	12.7	0.043	27	-20
Na_2O	5.60	250	0.53	9.3	0.033	14	-33
K_2O	5.70	200	0.42	11.4	0.040	13	-34

† At room temperature (300°K).

The next step is to compute similar enrichment factors for the combined oxides which make up rocks in the earth's crust.   When oxide separation factors are combined to give mineral separation factors, care must be taken to balance properly the oxygen atoms.   For example,

$$\text{Nepheline (Na}_2\text{Al}_2\text{Si}_4\text{O}_8) = 2\text{SiO}_2 + \text{Al}_2\text{O}_3 + \text{Na}_2\text{O}$$

$$\Delta^{\text{neph}}_{\text{qtz}} = \tfrac{4}{8}\,\Delta^{SiO_2}_{\text{qtz}} + \tfrac{3}{8}\,\Delta^{Al_2O_3}_{\text{qtz}} + \tfrac{1}{8}\,\Delta^{Na_2O}_{\text{qtz}}$$

$$= \tfrac{4}{8}(0) - \tfrac{3}{8}(5) - \tfrac{1}{8}(33)$$

$$= -\tfrac{48}{8} = -6.0$$

Values calculated in this way are given in Table 7.2.   As is shown in Fig. 7.3, the sequence of fractionations obtained in this way is consistent with the sequence of oxygen isotopic compositions found in igneous rocks.

TABLE 7.2

**Calculated Fractionations Relative to Quartz
for Important Rock-forming Minerals**

Mineral	Component Oxides	Δ_{qtz}^{min}†
Quartz	SiO_2	0.0
Sillimanite	$SiO_2 + Al_2O_3$	-3.0
Kyanite	$SiO_2 + Al_2O_3$	-3.0
Albite	$6SiO_2 + Al_2O_3 + Na_2O$	-3.0
Orthoclase	$6SiO_2 + Al_2O_3 + K_2O$	-3.1
Plagioclase	(albite + anorthite)	-3.5
Anorthite	$4SiO_2 + 2Al_2O_3 + 2CaO$	-4.1
Leucite	$4SiO_2 + Al_2O_3 + K_2O$	-4.1
Nepheline	$2SiO_2 + Al_2O_3 + Na_2O$	-6.0
Pyrope	$3SiO_2 + Al_2O_3 + 3MgO$	-5.0
Grossularite	$3SiO_2 + Al_2O_3 + 3CaO$	-5.2
Enstatite	$SiO_2 + MgO$	-5.0
Augite	$2SiO_2 + Al_2O_3 + CaO + MgO$	-5.3
Forsterite	$SiO_2 + 2MgO$	-7.5
Fayalite	$SiO_2 + 2FeO$	-10.0
Rutile	TiO_2	-6.0
Ilmenite	$TiO_2 + FeO$	-10.7
Hematite	Fe_2O_3	-14.0
Magnetite	$Fe_2O_3 + FeO$	-15.5

† At room temperature.

KINETIC FRACTIONATIONS

Isotope fractionation measurements taken during irreversible chemical reactions always show a preferential enrichment of the lighter isotope in the products of the reaction. S^{32} is enriched relative to S^{34} in the H_2S produced during bacterial reduction of $CaSO_4$. C^{12} is enriched relative to C^{13} in the organic molecules produced during photosynthesis. O^{16} is enriched relative to O^{18} in CO_2 and H_2O produced during the bacterial oxidation of organic debris.   The nature of this fractionation stems from the lower ground-state vibration frequency of the heavy isotope.   As shown in Fig. 7.1, its zero-point energy lies closer to the bottom of the potential-energy well than does that for the light isotope.   Hence more energy is required to destroy a molecule bearing the heavy isotope.   In Chap. 5 we demonstrated the great sensitivity of reaction rates to the

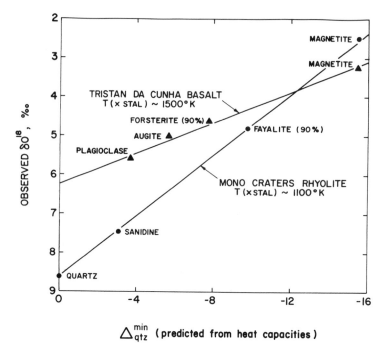

$\triangle^{min}_{qtz}$ (predicted from heat capacities)

Fig. 7.3. Comparison of relative degrees of O^{18} enrichment predicted by the heat-capacity model with those observed in the minerals separated from a rhyolite and from a basalt. The lower slope of the line connecting the basaltic minerals as compared with that for the rhyolitic minerals reflects the lower temperature of crystallization for the rhyolite. The fact that the points from a given rock lie along straight lines indicates that, despite the great oversimplification of the problem, the heat-capacity model adequately predicts the first-order differences in the degree of O^{18} enrichment in various oxide minerals. (*Data from Garlick.*)

magnitude of energy barriers. It is easily shown that, despite the small difference in ground-state energy levels produced by isotopic substitutions, they lead to significant rate differences. We showed that the rate, R, at which a reaction proceeded was proportional to $\exp(-E_B/RT)$, where E_B is the barrier energy. For the dissociation of a molecule the barrier energy is $N(\epsilon - \frac{1}{2}h\nu)$, where ϵ is the depth of the potential well and $\frac{1}{2}h\nu$ is the zero-point energy. The ratio of the rate for the heavy (R_H) to that for the light isotope (R_L) is

$$\frac{R_H}{R_L} = \frac{\exp\left[-N(\epsilon - \frac{1}{2}h\nu_H)/NkT\right]}{\exp\left[-N(\epsilon - \frac{1}{2}h\nu_L)/NkT\right]}$$

$$= \exp\left[-\frac{h}{2kT}(\nu_L - \nu_H)\right]$$

Since

$$\nu_H = \nu_L \sqrt{\frac{\mu_L}{\mu_H}}$$

$$\frac{R_H}{R_L} = \exp\left[-\frac{h\nu_L}{2kT}\left(1 - \sqrt{\frac{\mu_L}{\mu_H}}\right)\right]$$

For O_2 at room temperature

$$\frac{R_H}{R_L} = \exp\left(-\frac{(6.6 \times 10^{-27})(4.74 \times 10^{13})(0.058)}{(2)(1.4 \times 10^{-16})(300)}\right) \cong e^{-0.18} \cong 0.83$$

Thus molecules bearing two O^{16} atoms should react about 17 percent more rapidly than those bearing one O^{18} and one O^{16} atom.

The actual fractionation observed in natural irreversible processes depends not only on the difference in reaction rate for different isotopic species but also on the extent to which the parent reservoir is depleted. If, for example, the reactants are entirely used there can be no net fractionation. The product must have exactly the same composition as the reactants. Thus kinetic fractionations observed in nature are generally less than that predicted by the ratio of the reaction rates.

RAYLEIGH DISTILLATION

Any isotope reaction carried out in such a way that the products are isolated immediately after formation from the reactants will show a characteristic trend in isotopic composition. Examples of this type of process are the progressive formation and removal of raindrops from a cloud and the formation of crystals from a solution too cool to allow diffusive equilibrium between the crystal interior and the liquid. The isotopic composition of residual water vapor (or of the residual liquid) is a function of the fractionation factor between vapor and water droplets (or between the liquid and surface layer of the growing solid) and can be calculated as follows: Let A designate the amount of the species containing the major isotope and B the amount of that containing the trace isotope. The rate at which either species reacts is proportional to its amount; thus

$$dA = k_A A$$

and

$$dB = k_B B$$

Since the two isotopes react at slightly different rates, $k_A \neq k_B$. We will let the ratio of these two constants, k_B/k_A, be equal to α. Thus

$$\frac{dB}{dA} = \alpha\, \frac{B}{A}$$

Rewriting in integral form, we obtain

$$\int_{B^0}^{B} \frac{dB}{B} = \alpha \int_{A^0}^{A} \frac{dA}{A}$$

where A^0 and B^0 are the initial amounts of the two isotopic species. Integrating, the following is obtained:

$$\ln \frac{B}{B^0} = \alpha \ln \frac{A}{A^0}$$

or

$$\frac{B}{B^0} = \left(\frac{A}{A^0}\right)^{\alpha}$$

Dividing both sides by A/A^0 and rewriting, we have

$$\frac{B/A}{B^0/A^0} = \left(\frac{A}{A^0}\right)^{\alpha-1}$$

Since species B makes up only a trace fraction of the total $A + B$, the fraction of residual material, f, is equal to A/A^0. Hence

$$\frac{B/A}{B^0/A^0} = f^{\alpha-1}$$

Subtracting unity from both sides,

$$\frac{B/A - B^0/A^0}{B^0/A^0} = f^{\alpha-1} - 1$$

As the per mil change in isotopic composition, δ, is given by $(1000) \times (B/A - B^0/A^0)/(B^0/A^0)$

$$\delta = 1000(f^{\alpha-1} - 1)$$

A natural example of a Rayleigh distillation is the fractionation between the oxygen isotopes in water vapor in clouds and the rain-drops released from the cloud. O^{18} is enriched by 10 per mil in the droplets.   Hence the O^{18} is depleted at a rate 1.010 times faster than the O^{16} ($\alpha = k_{O^{18}}/k_{O^{16}} = 1.010$):

$$\delta = -1000(1 - f^{0.010})$$

The resulting depletion of the O^{18}/O^{16} ratio in the residual water vapor is given as a function of the fraction of original vapor remaining in the cloud (Fig. 7.4).   As the latter closely depends on the cloud temperature, the relationship between isotopic composition and cloud temperature is also given (Fig. 7.5).

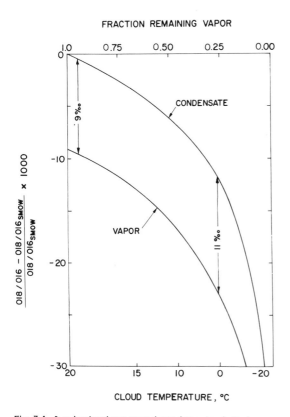

Fig. 7.4. $\delta_{O^{18}}$ in cloud vapor and condensate plotted as a function of the fraction of remaining vapor in the cloud for a Rayleigh process. The temperature of the cloud is shown on the lower axis. The increase in fractionation with decreasing temperature is taken into account. (*After Dansgaard, 1964.*) SMOW stands for standard mean ocean water.

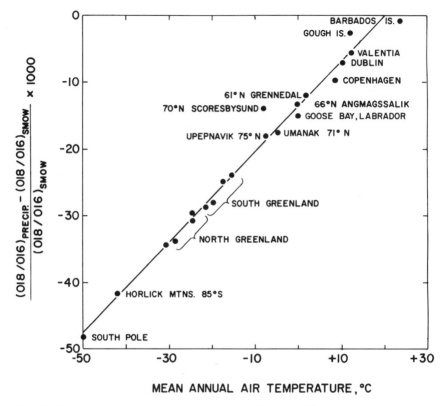

Fig. 7.5. Observed $\delta_{O^{18}}$ concentration in average annual precipitation as a function of mean annual air temperature. (*After Dansgaard, 1964.*)

SURFACE EQUILIBRIUM VERSUS INTERIOR EQUILIBRIUM

Because the atoms residing in the first few molecular layers of a crystal are not subject to the infinite series of regularly spaced electrostatic forces which act on interior atoms, they have somewhat different internal and free energies. Isotopic equilibria involving surface atoms may have different fractionations than those involving interior atoms. Thus the fractionation occurring in a natural process could depend on whether diffusion is fast enough to maintain equilibrium between atoms buried in the interiors of the solid phases present. For most reactions there will be a temperature above which this equilibrium will always be maintained and a temperature below which it will never be achieved. Between these two limits there exists a zone of uncertainty where the equilibrium may or may not have been achieved or even where it was partially achieved. Unfortunately, most metamorphic processes probably take place within this shadow zone; consequently, isotope separation observed in such rocks must be interpreted with great caution.

PROBLEMS

7.1 Consider a sediment composed of 50 mole percent detrital igneous quartz ($\delta = +10\ \%_0$) and 50 mole percent marine calcite ($\delta = +30\ \%_0$). Determine the oxygen isotopic composition of each mineral after metamorphic equilibration at 500°C ($\alpha^{SiO_2}_{CaCO_3} = 1.002$ at 500°C).

7.2 A sediment composed of 50 percent quartz and 50 percent calcite as in Prob. 7.1 is gradually decarbonated at 500°C to produce wollastonite. Apply the Rayleigh equation to calculate the composition of the rock after complete conversion. Use $\alpha^{CO_2}_{CaCO_3} = 1.007$, and assume that equilibrium fractionations among the condensed phases are negligible.

7.3 The O^{18} content of shells grown from seawater is larger than that in the water. If

$$\Delta^{CaCO_3}_{H_2O} = 30\ \%_0 \quad \text{at} \quad T = 25°C$$
$$= 34\ \%_0 \quad \text{at} \quad T = 5°C$$

(a) What is the temperature coefficient of the fractionation?

(b) Calculate the expected O^{18} content of surface-dwelling foraminifera at $T = 15°C$ and $T = 20°C$ with respect to standard mean ocean water (SMOW). If the 20°C sample were mixed with 10 percent bottom-dwelling forams ($T = 2°C$), what would the average O^{18} content be?

(c) The water currently locked up in the Greenland and Antarctic ice caps has 25 per mil less O^{18} than mean seawater. If during the maximum of the last glacial period the amount of additional ice equaled 4 percent of the water now in the oceans, and if this ice had the same O^{18} deficiency as the current ice caps, what average apparent temperature change would be induced in the average foram shell living in the glacial ocean, because of the change in isotopic composition of the residual water?

7.4 It is possible to separate U^{235} from U^{238} by successive gaseous diffusion of UF_6 through a series of small orifices. Each orifice is a separate stage. The natural U^{238}/U^{235} is 137.8. If $\alpha = 1.0004$ for each stage, how many stages will be required to produce (a) $U^{235}/U^{238} = 1.0$; (b) $U^{235}/U^{238} = 10.0$; (c) $U^{235}/U^{238} = 100.0$. Consider that only the light fraction from each stage passes on to the next stage.

7.5 The isotopic composition of the element Sr is measured in a mass spectrometer. The Sr is placed on a ribbon of refractory metal such as rhenium. The ribbon is then heated by passing a current through it to vaporize Sr as Sr^0. The Sr^0 is ionized to Sr^+ by passing over a filament at a temperature of 1400°C. The vaporization step occurs at a temperature of about 500°C in vacuo. Which of the two steps is most likely to produce an isotopic fractionation? If $(Sr^{86}/Sr^{87})_{vapor} = 1.002(Sr^{86}/Sr^{87})_{solid}$, calculate the change in isotopic composition of the residual solid (for Rayleigh distillation) as a function of the amount of solid remaining.

7.6 A small closed-basin lake receives 90 percent of its water from river inflow ($\delta_{O^{18}} = -16$) and the remaining 10 percent from direct rainfall ($\delta_{O^{18}} = -5$). This input is exactly balanced by evaporation from the lake surface. If the steady-state $\delta_{O^{18}}$ value for the lake water is -6, what is $\alpha = (O^{18}/O^{16})_{vapor}/(O^{18}/O^{16})_{lake}$ for the evaporating vapor?

REFERENCES

Dansgaard, W.: Stable Isotopes in Precipitation, *Tellus*, **16**: 436 (1964).

Garlick, G. D.: The Stable Isotopes of Oxygen, in K. H. Wedepohl (ed.), "Handbook of Geochemistry," vol. 2, sec. 8B, Springer-Verlag OHG, Heidelberg, 1969.

SUPPLEMENTARY READING

*Bottinga, Y.: Calculated Fractionation Factors for Carbon and Hydrogen Isotope Exchange in the System Calcite–Carbon Dioxide–Graphite–Methane–Hydrogen–Water Vapor, *Geochim. Cosmochim. Acta*, **33**: 49 (1969).

Clayton, R. N., B. F. Jones, and R. A. Berner: Isotope Studies of Dolomite Formation under Sedimentary Conditions, *Geochim. Cosmochim. Acta*, **32**: 415 (1968).

Craig, H.: Isotopic Composition and Origin of the Red Sea and Salton Sea Geothermal Brines, *Science*, **154**: 1544 (1966).

Deines, P., and D. P. Gold: The Change in Carbon and Oxygen Isotopic Composition during Contact Metamorphism of Trenton Limestone by the Mount Royal Pluton, *Geochim. Cosmochim. Acta*, **33**: 421 (1969).

Emiliani, C.: Pleistocene Temperatures, *J. Geol.*, **63**: 538 (1955).

Kemp, A. L. W., and H. G. Thode: The Mechanism of the Bacterial Reduction of Sulphate and of Sulphite from Isotope Fractionation Studies, *Geochim. Cosmochim. Acta*, **32**: 71 (1968).

Taylor, H. P., and S. Epstein: Relationship between O^{18}/O^{16} Ratios in Coexisting Minerals of Igneous and Metamorphic Rocks, *Geol. Soc. Am. Bull.*, **73**: 461 (1962).

chapter eight

Distribution of Trace Elements between Coexisting Phases

The elements O, Si, Al, Fe, Ca, Na, K, Mg, and Ti make up 99 percent of the earth's crust and mantle. The other 80 elements together contribute only 1 percent to the weight of the earth; these elements are collectively referred to as *trace elements*. Trace elements occasionally form minerals in which they are a major constituent, but more frequently a trace element occurs as an intruder in the minerals formed by the major elements. For example, all calcite contains a small amount of strontium. The strontium occupies lattice positions normally belonging to calcium ions. We may consider the carbonate mineral as consisting of a dilute solution of strontium carbonate in calcium carbonate. Whenever a trace element occurs as a minor constituent in a mineral or liquid phase, we may describe the behavior of the trace element by the laws of dilute solutions. We will first consider the behavior of dilute solutions of solids in liquids and of solids in solids and then examine several applications of trace-element distribution in natural systems.

THE LAWS OF DILUTE SOLUTIONS

In dealing with dilute solutions we have two problems to consider. First, what is the effect of a small amount of added trace material on the properties of the host liquid and, second, what is the effect of the host liquid on the properties of the dissolved trace element? We will first consider the change in the host's properties.

When a small amount of sugar is added to water, the solid phase disappears. The particles of solid sugar interact with the liquid water molecules, and a single phase, a dilute solution of sugar (the solute) in water (the solvent), results. If the solution is very dilute, the fugacity of the water will change only because its mole fraction is lowered, and not because of any interactions with the sugar molecules. Thus, for the solvent

$$f_1 = f_1^{\circ} x_1$$

where f_1 is the fugacity of the solvent in the dilute solution, f_1° is the fugacity of pure solvent at the same temperature and pressure, and x_1 is the mole fraction of solvent. The fugacity is proportional to the mole fraction because interactions between water molecules are so much more frequent than interactions between water and solute molecules. This result is an extension of Raoult's law which states that the vapor pressure of the solvent over a solution is proportional to the mole fraction of solvent. At low pressures, where fugacity and vapor pressure are identical, Raoult's law states

$$p_1 = p_1^{\circ} x_1$$

If only one solute is present, with mole fraction x_2,

$$p_1 = p_1^{\circ}(1 - x_2)$$

which may be rearranged as

$$\frac{p_1^{\circ} - p_1}{p_1^{\circ}} = x_2$$

This is the original form of Raoult's law. It states that the lowering of the vapor pressure of a solvent in a solution with respect to the vapor pressure of pure solvent is proportional to the mole fraction of solute. As the solution becomes more concentrated and solvent-solute interactions become important, the solvent will no longer follow Raoult's law. However, as we are considering trace-element properties and dilute solutions, Raoult's law will suffice.

Now let us turn to the second problem, the properties of the trace substance.   For solutes in dilute solutions, solute-solute interactions are negligible; solute-solvent interactions control the properties of the solute. In the limit of "infinite" dilution all solutes exhibit the same type of behavior.   This behavior is described by Henry's law, which states that the fugacity of a solute in an extremely dilute solution is proportional to the mole fraction of solute.   Thus, if the solute is again given the subscript 2,

$$f_2 = kx_2$$

where the constant of proportionality, k, is called the Henry's law constant.   This constant depends on the nature of both the solute and the solvent.

In very dilute solutions the mole fraction of solute is directly proportional to the concentration of solute.   A modified Henry's law constant, k', relates the fugacity to concentration:

$$f_2 = k'C_2$$

Since the solutions we will consider in this chapter are very dilute, Henry's law will be adequate.

ACTIVITY AND THE ACTIVITY COEFFICIENT

The fugacity of any substance is ultimately related to its vapor pressure. For solids, which have very low vapor pressures, the fugacity is extremely difficult to measure directly.   It is convenient to make use of a reference state with fugacity $f°$ and define the activity, a, of any substance as the ratio of its fugacity in the state of interest to its fugacity in the standard reference state:

$$a = \frac{f}{f°}$$

The following important consequences of this definition should be carefully noted.   First, the activity in the standard state is unity:

$$a° = \frac{f°}{f°} = 1$$

Second, the activity is directly related to the chemical potential:

$$RT \ln a = \mu - \mu°$$

where $\mu°$ is the chemical potential of the standard state. Since the activity is directly proportional to the fugacity, we have

$$d \ln a = d \ln f$$

One of the most useful properties of activity is that the reference state may be chosen to suit the problem at hand. In choosing a reference state, the main objective is to simplify the calculations. Consequently the standard state for gases differs from that for liquids and solids. Also, the choice for solvents differs from that for solutes. The following standard states are the ones generally chosen for low-pressure calculations:

1. Gases

The standard reference state for gases is taken from the hypothetical ideal gas state. The reference is chosen as the state at each temperature where an ideal gas would exhibit unit fugacity. Since the reference state is taken to be that for which the fugacity is unity, the activity of any gas is equal to its fugacity:

$$a = \frac{f}{f°} = \frac{f}{1}$$

Figure 8.1 shows a typical fugacity versus pressure curve for a real gas

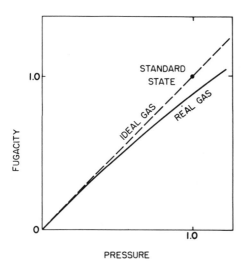

Fig. 8.1. Fugacity versus pressure for a real gas, showing the ideal gas reference state.

together with the ideal gas reference state.   We have shown in Chap. 6 that the fugacity of a real gas approaches the pressure as the pressure becomes very small.   In the limit of very low pressure, the activity of a real gas is therefore equal to its pressure.

2. Solvents

The standard state for a solvent is the pure liquid or solid at the same temperature and 1 atm pressure.   Thus the activity approaches the mole fraction as the mole fraction approaches unity:

$$\frac{a}{x} \to 1 \qquad \text{as} \qquad x \to 1$$

3. Solutes

For dilute solutions it is convenient to choose the standard state so that in the limit of infinite dilution the activity of a solute will equal its mole fraction.   Thus

$$a \to x \qquad \text{as} \qquad x \to 0$$

In order to find a reference state to fit our demands, we must again resort to a hypothetical standard state.   Since the fugacity of the reference state must satisfy the identity

$$f^\circ = \frac{f}{a}$$

and since we require that $a \to x$ as $x \to 0$, we find that

$$f^\circ = \lim_{x \to 0} \frac{f}{x}$$

The limit of the ratio f/x as a solution becomes very dilute is simply the Henry's law constant, as shown earlier in this chapter.   If the state of infinite dilution itself were taken to be the reference state, the standard chemical potential, μ°, would be minus infinity.   This is clearly an undesirable result.   To avoid this complication the Henry's law line is extrapolated from zero mole fraction to unit mole fraction.   The reference state is then taken as a solute with unit mole fraction, yet which obeys Henry's law.   This state, which is clearly hypothetical, is illustrated in Fig. 8.2.

In most cases we wish to express the behavior of a solute in terms of ideal behavior plus departures from ideality. In an ideal solution Henry's law holds and the activity of the solute is equal to its mole fraction. Thus, for an ideal solution

$$\frac{a}{x} = 1$$

The departure of the ratio, a/x, from unity is then a measure of nonideality in the behavior of a solute. This ratio is called the activity coefficient, γ:

$$\gamma = \frac{a}{x}$$

Thus, for a solute which behaves ideally, γ is equal to unity.

THE STANDARD STATE AT HIGH PRESSURE

The chemical potential of a solute in a solution is a function of the temperature, pressure, and composition of the solution. For the preceding discussion of activity we considered only solutions at 1 atm pressure. At the constant pressure of 1 atm the chemical potential of a solute will change as the temperature and composition of the solution change.

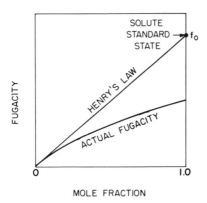

Fig. 8.2. Reference state for solutes in dilute solutions. The Henry's law line is extrapolated to unit mole fraction. The reference state is taken to be the hypothetical state at unit mole fraction.

Since the chemical potential of the reference state, μ°, depends on temperature, we have that

$$\mu_i(T,x,P = 1 \text{ atm}) = \mu_i^\circ(T,P = 1 \text{ atm}) + RT \ln \gamma_i x_i$$

In order to treat solutions at elevated pressures we must include the effect of pressure on the chemical potential.   Most American textbooks do this in a rather circuitous fashion by introducing another variable under the logarithm.   European textbooks use a more convenient approach which includes the effect of pressure in the evaluation of the chemical potential of the reference state. This reserves the logarithm term for dealing with the effects of concentration and nonideality.   Thus

$$\mu_i(T,x,P) = \mu_i^*(T,P) + RT \ln \gamma_i x_i$$

To avoid confusion we will use μ° for reference states at 1 atm, and μ^* for reference states at arbitrary pressures.

DISTRIBUTION OF A TRACE ELEMENT BETWEEN TWO SOLIDS

Let us consider the distribution of a trace element between two coexisting phases with which it forms dilute solutions.   The chemical potentials of the trace element in phases I and II are given by

$$\mu_I = \mu_I^* + RT \ln a_I$$
$$\mu_{II} = \mu_{II}^* + RT \ln a_{II}$$

The quantities μ_I^* and μ_{II}^* are the chemical potential of the trace element in its standard state (see Fig. 8.2) at the temperature and pressure of interest.   At equilibrium we must have

$$\mu_I = \mu_{II}$$

Since μ_I^* and μ_{II}^* are constants at fixed T and P, it follows that

$$\frac{a_I}{a_{II}} = \text{const} \equiv K'$$

This constant is sometimes called the distribution coefficient.   For dilute solutions which follow Henry's law we may replace the activity ratio by

$$\frac{k_I x_I}{k_{II} x_{II}} = \text{const}$$

or

$$\frac{x_I}{x_{II}} = K$$

It can be shown (see Appendix I) that the distribution coefficient varies with temperature according to the relation

$$\left(\frac{\partial \ln K}{\partial T}\right)_P = \frac{\Delta H}{RT^2}$$

and

$$K = -\frac{\Delta H}{RT} + B$$

where ΔH represents the difference between the heats of solution for the trace element in phases I and II, and B is a constant of integration.

An interesting application of these concepts has been made by Häkli and Wright. They measured the distribution of nickel (present in trace quantities) between the phases olivine, clinopyroxene, and glass for samples collected from the Makaopuhi lava lake in Hawaii. They hoped to use the distribution of Ni between coexisting phases to determine the temperature of crystallization of igneous bodies.

In order to use Ni distribution as a geothermometer, the phases most sensitive to temperature changes must be identified and the temperature scale must be calibrated. It was hoped that the work on the Hawaiian lava lake would accomplish this. To avoid the complications involved by changes in composition of the olivine and pyroxene phases, only samples of constant bulk composition were used.

Samples were taken from the lava lake after varying degrees of crystallization had occurred. The temperature was measured at the time of collection and then the sample was quenched (cooled rapidly) to convert any liquid present to glass. The Ni content of the phases olivine, clinopyroxene, and glass were measured, and the distribution coefficients were calculated. The results of their measurements are given in Table 8.1. The distribution of Ni between the mineral pair olivine-clinopyroxene appears to be fairly sensitive to temperature, particularly in the high-temperature range.

By plotting the distribution coefficients against $1/T$, a value of ΔH

TABLE 8.1

Distribution of Nickel between Coexisting Olivine, Pyroxene, and Glass for the Makaopuhi Lava Lake†

Sample	Temp, °C	Ol^{Ni}, ppm	Py^{Ni}, ppm	Gl^{Ni}, ppm	Ol^{Ni}/Gl^{Ni}	Py^{Ni}/Gl^{Ni}	Ol^{Ni}/Py^{Ni}
I	1160	1555	255	115	13.5	2.22	6.10
II	1120	1310	245	87	15.1	2.82	5.35
III	1075	955	240	60	15.9	4.00	3.98
IV	1070	935	235	57	16.4	4.12	3.98
V	1050	840	220	50	16.8	4.40	3.82

† Data taken from T. A. Häkli and T. L. Wright, The Fractionation of Nickel between Olivine and Augite as a Geothermometer, *Geochim. Cosmochim. Acta*, **31**: 877–884 (1967).

for each of the pairs may be found.   This is shown in Fig. 8.3 and Table 8.2.   The straight line may be extrapolated to lower temperatures to provide estimates of the partitioning of Ni between olivine and pyroxene in intrusive bodies.   However, care must be taken when extrapolating data of this type since changes in ΔH with temperature may become important.   Also, for intrusive bodies the effects of pressure on the chemical potential may become important.

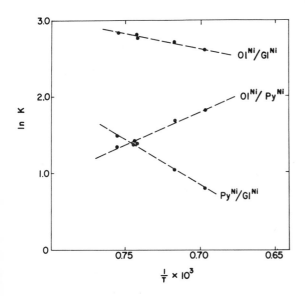

Fig. 8.3. Distribution coefficient for Ni between coexisting phases plotted as a function of $1/T$. (*Data taken from Häkli and Wright.*)

TABLE 8.2

Parameters Needed to Calculate the Variation of Nickel Distribution Coefficient as a Function of Temperature[†]

	ΔH, cal/mole	B
Ol^{Ni}/Gl^{Ni}	$-7,400$	0.03
Py^{Ni}/Gl^{Ni}	$-24,800$	-7.85
Ol^{Ni}/Py^{Ni}	$16,800$	$+7.65$

† Data taken from T. A. Häkli and T. L. Wright, The Fractionation of Nickel between Olivine and Augite as a Geothermometer, *Geochim. Cosmochim. Acta*, **31**: 877–884 (1967).

FACTORS WHICH INFLUENCE DISTRIBUTION COEFFICIENTS

A great deal of data is available on the mutual solubility of the alkali halides under various conditions.   By examining the distribution coefficients for different cation pairs with a constant anion we may gain some insight into the factors which influence distribution coefficients.   Table 8.3 gives the ratio of the substituting ion to the major ion in the crystal divided by the same ratio for the solution or melt.   For example, the distribution coefficient for RbCl between a crystal of KCl and a solution of KCl (or KCl melt at high temperatures) would be

$$D_{KCl}^{RbCl} = \frac{(RbCl/KCl)_{cryst}}{(RbCl/KCl)_{sol}}$$

TABLE 8.3

Distribution Coefficients for Alkali Halides between Crystals and Aqueous Solution at 40°C and between Crystals and Melts

Solute	Radius of Cation	Distribution Coefficients[†]			
		KCl and Aqueous Solution, 40°C	KCl Crystals and Melt, 775°C	NaCl Crystals and Melt, 800°C	RbCl Crystals and Melt, 715°C
NaCl	0.97	1.3×10^{-3}	0.37	. . . .	0.21
KCl	1.33	. . . .	. . . .	0.21	1.01
RbCl	1.47	0.10	0.65	0.012	
CsCl	1.67	1.14×10^{-4}	0.27	0.005	0.56

† Data taken from Jörg Reichert, Verteilung anorganischer Fremdionen bei der Kristallisation von Alkalichlouden, *Contrib. Mineral. Petrol.*, **13**: 134–160 (1966) and Hans Hartmut Schock, Bestimmung sehr kleiner Verteilungskoeffizienten von Cs, Na und Ba zwischen Lösung und KCl-Einkristallen mit Hilfe radioaktiver Isotope, *Contrib. Mineral. Petrol.*, **13**: 161–180 (1966).

By comparing the data for KCl crystals grown at low temperatures with those for crystals grown from melts, it is obvious that ionic substitution is much more prevalent at high temperatures.   The increased solubility at high temperatures stems from two causes, both of which are related to the vibration of atoms within the lattice.   First, the increased vibrational amplitude and asymmetry at high temperatures result in expansion of the crystal structure, allowing easier substitution of the larger Rb and Cs ions for the smaller K ion.   Second, the increased vibration amplitude tends to make smaller ions, such as Na, appear larger because the extremes of vibration carry the ions farther from their equilibrium positions.   Although the effects of ionic size are greatly reduced at high temperature, they are not completely removed.   The ion most similar to the host in size (Rb in this case) still has the greatest solubility in KCl crystals grown from a melt at 775°C.

As the difference in size between the major cation and the substituting ion becomes large, the control of size becomes important even at high temperatures.   The effect is particularly important for a large ion substituting into the lattice of a small ion.   Thus the distribution coefficient for CsCl in NaCl is very small even in crystals grown from a NaCl melt.   For a similar difference in ionic radius where one ion is larger and one smaller than the major ion, such as the case for K and Cs substituting in RbCl, the smaller ion will have the larger distribution coefficient.

BROMINE AND THE DEPOSITIONAL HISTORY OF EVAPORITES

Seawater has an average salt content of 3.5 percent by weight.   If seawater is placed in a beaker and the water is allowed to evaporate, the salinity of the solution will obviously increase as the water content decreases.   Eventually a very concentrated salt soluton, called a brine, will result.   The first salts to precipitate from the brine will be aragonite and gypsum.   When the volume of the brine reaches approximately 13 percent of the volume of the original seawater solution, halite (NaCl) will start to precipitate.   If evaporation is continued further, sylvite will precipitate when the volume of brine is 5 percent of the original volume.   Other complex salts such as $KMgCl_3 \cdot 6H_2O$ (carnallite) and $MgCl_2 \cdot 6H_2O$ (bischofite) may also form in cases of extreme evaporation.

Seawater contains a small amount of bromine in the form of Br^- (bromide ion).   Since the bromide concentration is low, no pure bromide minerals form.   All the bromide is taken up as a dispersed element in the chloride salts.   The concentrations of bromide, even in brines, are low enough so that Henry's law is obeyed.

Braitsch and Herrmann measured the distribution coefficient for Br ion between seawater and NaCl.   They found that at 25°C the value of the distribution coefficient at the beginning of halite precipitation was

$$K_D = \frac{[\text{Br}^-]_{\text{NaCl}}}{[\text{Br}^-]_{\text{seawater}}} = 0.15$$

This means that the bromide content of the brine increases more rapidly than the salinity of the brine once halite begins to precipitate.   The value of K_D remains constant up to very high bromide concentrations. (It does depend on the $MgCl_2$ content of the solution at high salinities.) If we find a halite bed in a sequence of evaporite salts we may use the bromide content in the NaCl to infer the salinity of the brine from which the halite formed.

This might at first seem a trivial result; however, the history of evaporite basins can be very complicated.   For instance, in some places very thick sequences of evaporites are found.   The thicknesses are so great that simple evaporation of a closed basin of water is thoroughly

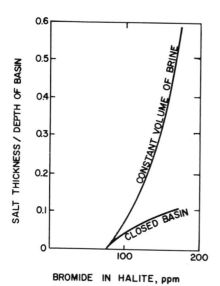

Fig. 8.4. Bromine content in halite plotted as a function of thickness of the salt deposited from a basin for two cases of water balance. The upper curve is for a case where the inflow of seawater just balances the loss due to evaporation. The lower curve is for simple evaporation of a closed basin. (*After Holser.*)

inadequate to explain the accumulated salts.   By examining the bromide content of NaCl and inferring the salinity of the brine at the time of deposition, we may decide whether the basin was filled and then evaporated many times or whether filling and evaporation were continuously going on.

Figure 8.4 shows two possible types of behavior for evaporite basins. The lower curve is for simple evaporation in a closed basin.   The upper curve is for the case where inflow of seawater just matches the loss of water by evaporation.   This produces a brine of constant volume with gradually increasing salinity.   For the constant-volume, "semi-open" basin, a much thicker sequence of salts will result.

The salt sequence found near Stassfurt, Germany, known as the Zechstein series, shows the stages through which many evaporite basins pass.   The bottom portion of the sequence, illustrated in Fig. 8.5, shows the initial buildup of salinity in the brine, resulting in halite formation. A large thickness of halite was precipitated under conditions which

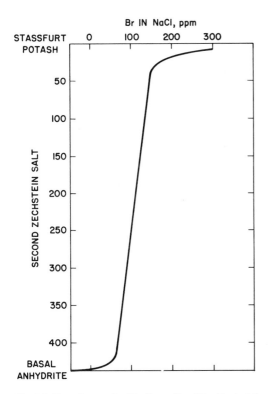

Fig. 8.5. Bromine content in the salts of the Zechstein series. (*After Holser.*)

allowed inflow of seawater to feed the basin. The inflow of seawater must have exceeded water loss by evaporation since the increase in bromide in the NaCl is slower than that predicted by the constant-volume model. The top part of the section indicates a rapid change to closed-basin conditions, resulting in high salinity and ultimately in precipitation of KCl, sylvite.

PROBLEMS

8.1 The element A is present initially in a silicate melt at the concentration a_0. The melt begins to crystallize in such a way that once formed the crystals do not react with the remaining liquid. If the ratio of the concentration of A in the newly formed crystals, a_{cryst}, to that in the residual liquid, a_{liq}, is always K, prove that the ratio of the concentration of A in the residual liquid to that in the starting melt is given by

$$\frac{a_{liq}}{a_0} = F^{K-1}$$

where F is the fraction of liquid remaining. This is the crystallization equivalent of Rayleigh distillation.

8.2 A zoned clinopyroxene in a basaltic rock has a chromium content of 5000 ppm at its center and 500 ppm at its edge. Assume that the distribution coefficient for chromium between liquid and pyroxene is $K_{cpx}^{liq} = 0.20$. Calculate the change in the amount of melt during the period over which the crystal grew.

8.3 Assume the following distribution coefficients between liquid and bulk solid:

$$\text{Uranium: } K_{sol}^{liq} = 5.0$$
$$\text{Thorium: } K_{sol}^{liq} = 6.0$$

Starting from a solid with U = 0.1 ppm and Th = 0.3 ppm, calculate the concentrations of U and Th in a partial melt and residual solid for cases where 1, 2, 5, 10, and 20 percent of the solid is melted (*a*) maintaining complete equilibrium between liquid and residual solid; (*b*) under conditions of Rayleigh distillation where liquid is removed from contact with the solids as soon as it is formed; (*c*) where the solid and melt do not equilibrate. Which process produces the largest change in the Th/U ratio? Which process produces the greatest enrichment of U and Th into the liquid?

*8.4 Assume that the distribution coefficients for K in the plagioclases are

$$\text{Albite: } K_{\text{liq}}^{\text{ab}} = 0.9$$
$$\text{Anorthite: } K_{\text{liq}}^{\text{an}} = 0.1$$

Starting from a liquid containing 1% $KAlSi_3O_8$, by varying the $NaAlSi_3O_8/CaAl_2Si_2O_8$ ratio in the liquid calculate the K content of the first feldspars to crystallize as a function of the plagioclase composition. Assume that the distribution coefficients do not vary with temperature.

REFERENCES

Braitsch, Otto, and Albert Günter Herrmann: Zur Geochemie des Broms in salenaren Sedimenteno.   Teil I, Experimentelle Bestimmung der Br-Verteilung in verschiedenen natürlichen Salzsystemen, *Geochim. Cosmochim. Acta*, **27**: 361–391 (1963).

Häkli, T. A., and T. L. Wright: The Fractionation of Nickel between Olivine and Augite as a Geothermometer, *Geochim. Cosmochim. Acta*, **31**: 877–884 (1967).

Holser, William T.: Bromide Geochemistry of Salt Rocks, in Jon L. Rau (ed.), "Second Symposium on Salt," pp. 248–275, The Northern Ohio Geological Society, Inc., Cleveland, Ohio, 1966.

———: After Gunter Schulze, Stratigraphische und genetische Deutung der Bromvertielung in den mittel deutschen Steinsalzlagern des Zechsteins, *Freiberger Forschungsh.*, **C83**.

SUPPLEMENTARY READING

Gast, P. W.: Trace Element Fractionation and the Origin of Tholeiitic and Alkaline Magma Types, *Geochim. Cosmochim. Acta*, **32**: 1057 (1968).

Greenland, L. P., D. Gottfried, and R. I. Tilling: Distribution of Manganese between Coexisting Biotite and Hornblende in Plutonic Rocks, *Geochim. Cosmochim. Acta*, **32**: 1149 (1968).

Haskin, L. A., and M. A. Haskin: Rare-earth Elements in the Skaergaard Intrusion, *Geochim. Cosmochim. Acta*, **32**: 433 (1968).

Hollister, L. S.: Garnet Zoning: An Interpretation Based on the Rayleigh Fractionation Model, *Science*, **154**: 1647 (1966).

Kinsman, D. J. J., and H. D. Holland: The Co-precipitation of Cations with $CaCO_3$. IV, The Co-Precipitation of Sr^{2+} with Aragonite between 16° and 96°C, *Geochim. Cosmochim. Acta*, **33**: 1 (1969).

Schwarcz, H. P.: The Effect of Crystal Field Stabilization on the Distribution of Transition Metals between Metamorphic Minerals, *Geochim. Cosmochim. Acta*, **31**: 503 (1967).

Whittaker, E. J. W.: Factors Affecting Element Ratios in the Crystallization of Minerals, *Geochim. Cosmochim. Acta*, **31**: 2275 (1967).

chapter nine **Solid-state Mineral Transformations**

The problem of discovering what chemical reactions take place in the earth's mantle occupies the attention of a large number of geophysicists and geochemists. From seismic data it is clear that the earth's mantle increases its density more rapidly with depth than would be expected from the compressibilities of oxide minerals. This excessively rapid increase could represent either the rearrangement of the atoms present into new, more compact crystalline phases or increase in the FeO or Fe content of mantle material. Thus solid-solid phase transformations have become a subject of considerable interest to the geophysicist.

SOLID–SOLID PHASE TRANSFORMATIONS

Many chemical compounds can exist in more than one crystalline form. Such substances are said to display polymorphism. For example, six polymorphs of SiO_2 are found in rocks exposed at the surface of the earth. These polymorphic forms are low quartz (α), high quartz (β), tridymite, cristobalite, coesite, and stishovite. Each is characterized by a different three-dimensional array of Si and O atoms. Silica glass is also found.

Of these seven solid phases, only one is stable at room temperature and pressure; the other phases are metastable.   The stable phase is the one with the lowest free energy under these conditions.   This phase is low quartz.   The occurrence of six other solid forms of silica in surface rocks reflects the extremely low rates of reaction at surface temperatures.

If the free energy of each of the phases is determined for other temperature and pressure combinations, not only will the values be different from those at room temperature and pressure, but also their order will change.   At 1 atm and 600°C high quartz will have the lowest free energy, whereas at 30,000 atm and 600°C coesite will be the stable phase.

By determining the relative free energies of these silica phases over a wide range of temperatures and pressures, the fields of stability of each phase can be mapped.   A phase diagram constructed in this manner for silica is shown in Fig. 9.1.   In addition to the fields for five of the seven solids the field for liquid silica is shown.   Stishovite forms at higher pressures than shown in the diagram.   Silica glass is the only phase which never possesses the lowest relative free energy.   It is metastable at all temperatures and pressures.

The boundaries separating the stability field of any two phases are smooth curves.   They constitute the locus of P, T points at which the

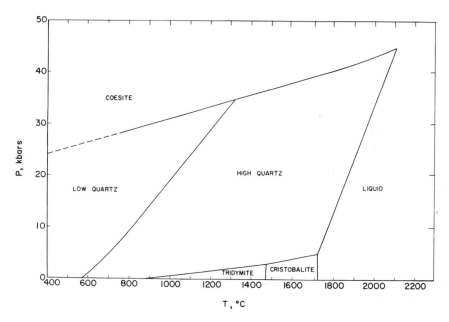

Fig. 9.1. Phase diagram for SiO$_2$, showing the stability fields of the polymorphic forms of SiO$_2$ and liquid SiO$_2$. (*Data from Boyd and England.*)

two phases have identical free energies.   Thus conversion of one form to the other at such points will not alter the entropy of the universe. The entropy change for the substance, ΔS, will be exactly balanced by a corresponding entropy change for the surroundings, $\Delta H/T$.

These phase boundaries will meet at so-called triple points where three phases can coexist at equilibrium.   Four phases of a single compound can never coexist at equilibrium.

The phase rule introduced by Gibbs can be used to analyze most simple systems.   If f is the number of variables required to uniquely define the system, C the number of independent components (or compounds) making up the system, and P the number of phases present, then at equilibrium

$$f = C + 2 - P$$

The silica system is a one-component system, and so

$$f = 3 - P$$

If only one phase, such as high quartz, is present at equilibrium, the state of the system is not defined unless both its temperature and pressure are given ($f = 3 - 1 = 2$).   If high quartz and cristobalite are costable, the state of the system is fixed by giving either the temperature or the pressure ($f = 3 - 2 = 1$).   If high quartz, cristobalite, and liquid SiO_2 coexist at equilibrium, there must be a triple point

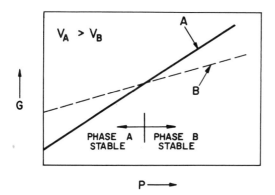

Fig. 9.2. Change in free energy with pressure for two phases of the same compound.   At low pressures phase A has the lowest free energy and is thus the stable phase.   As the pressure is increased, the free energy of phase A increases more rapidly than that of phase B, because of the larger volume of phase A. Eventually, the free energy curves cross, and phase B becomes stable.

($f = 3 - 3 = 0$).   The coexistence of four or more phases leads to a negative value for f and is not possible.   For one-component systems the phase rule merely states the obvious; however, this rule becomes extremely useful in dealing with multicomponent chemical systems.

The arrangement of stability fields for the phases of a given compound is closely related to their respective volumes and entropies.   We have already shown that

$$\left(\frac{\partial G}{\partial T}\right)_P = -S$$

and

$$\left(\frac{\partial G}{\partial P}\right)_T = V$$

These relationships require that along any traverse across a phase diagram with increasing temperature at constant pressure, phases of successively *higher* entropy will be encountered, and along any constant-temperature traverse with increasing pressure, phases with successively *lower* volume will be encountered.   Figures 9.2 and 9.3 make clear the reason for these progressions.   Conservation of volume becomes more and more important with increasing pressure, and randomness more and more important with increasing temperature.

The nature of the boundaries separating the stability fields of phases reveals some important relationships between entropy and volume.   In

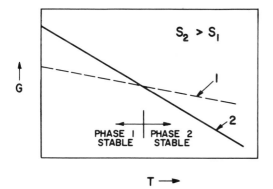

Fig. 9.3. Change in free energy with temperature for two phases of the same compound. At low temperatures phase 1 has the lowest free energy. However, phase 2 has the larger entropy and, as the temperature increases, its free energy drops more rapidly than that of phase 1.   At high temperatures, after the two free energy curves cross, phase 2 will be stable.

order to appreciate this we must derive the Clapeyron equation which relates the slope of a phase boundary to the volume and the entropy change for the transformation represented by the boundary.

Consider two closely spaced points on a phase boundary. For each the free energy of phase A must equal that of phase B. Hence the change in the free energy of phase A in going from one point to the other will be equal to that for phase B, or

$$dG_A = dG_B$$

We have already shown that

$$dG = V\, dP - S\, dT$$

Hence

$$V_A\, dP - S_A\, dT = V_B\, dP - S_B\, dT$$

Rearranging,

$$\left(\frac{dP}{dT}\right)_{\text{P.B.}} = \frac{S_B - S_A}{V_B - V_A} = \frac{\Delta S}{\Delta V}$$

The slope of the boundary must equal the ratio of the entropy change to the volume change for the reaction.

Phase boundaries generally have positive slopes. Thus phases with larger volumes generally have higher degrees of randomness. Although this relationship is intuitively clear for transitions involving liquids and vapors, it is not obvious that it should apply to solid-solid transitions. For example, the various crystalline forms of silica all have identical entropies at absolute zero. If the phase with largest volume (i.e., tridymite) is to have highest entropy at some finite temperature, T, then its integral

$$\int_0^T \frac{C_P}{T}\, dT$$

must be the greatest. This can be true only if tridymite (the least dense form of silica) also has the highest heat capacity.

Before explaining why more open crystal structures have higher heat capacities than their more dense counterparts, let us consider another aspect of solid-solid phase boundaries. Above room temperature these boundaries generally are almost straight lines. This implies that $\Delta S/\Delta V$ remains constant with increasing temperature. Because of low compressibilities and low coefficients of thermal expansion, the volume of solids changes very slowly with changing P and T. If the ΔV for the reaction of interest is greater than a few percent of the volume of the phases involved, the ΔV will not change greatly between 300 and 700°K. This is shown in Fig. 9.4 for the reaction calcite-aragonite.

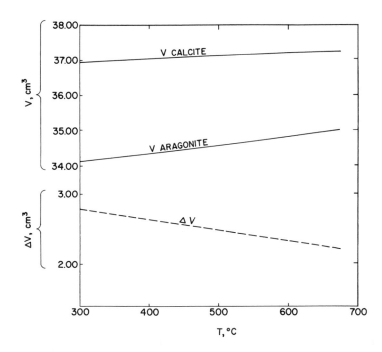

Fig. 9.4. Volumes of calcite and aragonite, and the difference in volume between the two phases, as a function of temperature.

If ΔV does not change appreciably along the phase boundary, the boundary can be a straight line only if ΔS does not change (above room temperature). As shown in Chap. 4, the entropy change at constant pressure is

$$\Delta S = \int_0^T \frac{\Delta C_P}{T} \, dT$$

A plot of $\Delta C_P/T$ versus T for the polymorphic pair sillimanite and kyanite is shown in Fig. 9.5. Clearly the major contribution to the entropy difference is generated below $300°K$. This reflects both the convergence of the heat capacities of the two phases with increasing temperature and the fact that heat added at high temperature affects the entropy far less than that added at low temperatures. The additivity of room-temperature heat capacities discussed in Chap. 3 demonstrates that for oxide reactions in general ΔC_P is small above room temperature.

Figure 9.5 also sheds some light on the question raised regarding the sense of phase boundaries. The convergence of the heat capacities of sillimanite and kyanite with increasing temperature indicates that the high-frequency portion of the vibrational spectrum must be nearly identical for the two polymorphs and that the low-frequency portion must be very different.   As the temperature is raised, the low-frequency vibrations become fully activated and their contributions to the heat capacity become constant.

The high-frequency modes of vibration in a crystal reflect the properties of individual anion-cation bonds.   They are therefore not particularly sensitive to the atomic arrangement.   On the other hand, the low-frequency modes involve groups of atoms.   They are, therefore, very sensitive to the particular atomic arrangement.

In order to understand the nature of the low-frequency differences, let us recall that the compressibility of an oxide decreases as it becomes more compact.   The compressibility of a solid is a measure of its rigidity, and its rigidity controls the natural acoustical vibration frequencies. Thus the more compact the oxide, the higher are its acoustical vibrations and the lower is its low-temperature heat capacity.   The positive slope of solid-solid phase boundaries is then a natural consequence of the relationship between volume per atom and bulk modulus shown in Chap. 2.

It should be emphasized that the generality that solid-solid phase boundaries should be positive and straight applies only to those transformations where the volume change is reasonably large (more than about 1 percent; the actual value depends on the relative compressibilities and thermal expansions for the two phases).   For reactions involving smaller changes in density the changes in ΔV and in ΔS which occur along the

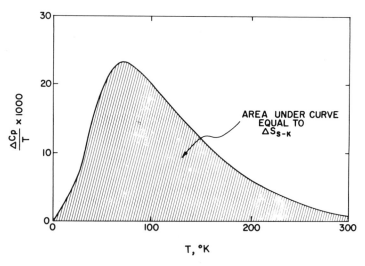

Fig. 9.5. Plot of $\Delta C_P/T$ versus temperature for the polymorphic pair sillimanite-kyanite.

boundary are much more important and could lead to considerable curvature. Also the relationship between compactness and the vibration spectrum is not so perfect as to rule out an inverse relationship between volume change and entropy change in these cases (i.e., negative slopes for phase boundaries). In Table 9.1 are listed examples of solid-solid reactions which may take place in the earth's crust and mantle. For each reaction the appropriate entropy data (as computed from measured heat capacities) are given. The differences in entropy for the reaction computed in this way are compared with those based on phase-boundary slopes (determined experimentally) and volume differences (based on densities computed from x-ray diffraction data). Hence the entropy change based on

$$\Delta S = \int_0^T \frac{\Delta C_P}{T} \, dT$$

is compared with the ΔS based on

$$\Delta S = \Delta V \left(\frac{dP}{dT}\right)_{\text{P.B.}}$$

Significant differences are not uncommon, especially as ΔS becomes a small fraction of the total entropy.

Perhaps the most interesting feature of these results is that all reactions for which ΔV is greater than 5 percent of the total volume of the reactants have phase-boundary slopes of 18 ± 6 atm/deg. This constancy suggests that the relationship between volume change and entropy change is more than just similarity in sense. The entropy change appears to be roughly proportional to the volume change. This is not unreasonable in light of the above discussion.

So far we have restricted the discussion to the slopes for solid-solid-reaction phase boundaries. Nothing has been said regarding their location in a pressure versus temperature diagram. For example, at 25°C at what pressure do calcite and aragonite have the same free energy? Since

$$G^P_{\text{calcite}} = G^{\text{std}}_{\text{calcite}} + \int_1^P V_C \, dP$$

$$G^P_{\text{aragonite}} = G^{\text{std}}_{\text{aragonite}} + \int_1^P V_A \, dP$$

then for $G^P_{\text{aragonite}} = G^P_{\text{calcite}}$,

$$\Delta G^{\text{std}} = - \int_1^P \Delta V \, dP$$

TABLE 9.1

Volume Change and Phase-boundary Slope for Several Solid-Solid Reactions. The Entropy Change Predicted by the Clapeyron Equation Is Compared with the Entropy Change Predicted from Heat-capacity Data

Reaction	ΔV, cm³/mole	dP/dT, bars/deg	ΔS,† cal/deg·mole	ΔS,‡ cal/deg·mole	$\Delta S\ddagger/S_{reac}$	$\Delta V/V_{reac}$
Graphite → diamond C C	−1.88	24.0	−1.09	−0.78	−0.57	−0.35
Calcite → aragonite $CaCO_3$ $CaCO_3$	−2.78	16.4	−1.10	−1.00	−0.046	−0.075
Sillimanite → kyanite Al_2SiO_5 Al_2SiO_5	−5.80	13.0	−1.82	−2.95	−0.128	−0.118
Quartz → coesite SiO_2 SiO_2	−2.05	11.06	−0.55	....	....	−0.090
Nepheline + albite → 2 Jadeite $NaAlSiO_4$ $NaAlSi_3O_8$ $NaAlSi_2O_6$	−32.42	17.9	−14.0	−14.7	−0.188	−0.21
Albite → jadeite + quartz $NaAlSi_3O_8$ $NaAlSi_2O_6$ SiO_2	−16.54	19.8	−7.9	−7.41	−0.150	−0.16

† Calculated from phase-boundary slope.
‡ Calculated from heat-capacity data.

Assuming ΔV to remain nearly constant with P and that $P_{equil} \gg 1$,

$$P^{25°C}_{equil} = -\frac{\Delta G^{std}}{\Delta V^{std}}$$

Since $\Delta G = \Delta H - T\,\Delta S$,

$$P^{25°C}_{equil} = -\frac{\Delta H^{std}}{\Delta V^{std}} + (298)\frac{\Delta S^{std}}{\Delta V^{std}} = \frac{-48 + 298(1)}{2.8 \times 10^{-3}}\frac{1}{24.2} = 3700 \text{ atm}$$

Further we have shown that $\Delta S^{std}/\Delta V^{std}$ is nearly equal to the slope of the phase boundary, so that

$$P^{T}_{equil} = P^{25°C}_{equil} + (T - 298)\frac{\Delta S^{std}}{\Delta V^{std}}$$

A phase boundary can be determined in two different ways. It can be directly fixed by experimentally determining which phases are stable at a series of P, T conditions. Alternatively, heats of solution of the various minerals can be measured in a calorimeter at 1 atm, 25°C. The difference between the heat of solution of the reactants and of the products is equal to the ΔH^{std} for the reaction. Enthalpy differences for the reactions mentioned above are shown in Table 9.2 along with values of P_{equil}.

TABLE 9.2

Enthalpy Differences and Equilibrium Pressure at Room Temperature for Several Solid-Solid Reactions

Reaction	ΔH^{std}, kcal/mole	$P^{25°C}_{equil}$, atm
Graphite $\rightarrow$ diamond	0.453	15,100
Calcite $\rightarrow$ aragonite	-0.048	3,720
Sillimanite $\rightarrow$ kyanite†	-0.266	4,370
Nepheline + albite $\rightarrow$ jadeite‡	-7.0	$-3,340$
Albite $\rightarrow$ jadeite + quartz‡	-0.5	4,270

† David R. Waldbaum, Thermodynamic Properties of Mullite, Andalusite, Kyanite and Sillimanite, *Am. Mineralogist*, **50**: 186–195 (1965).
‡ F. C. Kracek, N. J. Neuvonen, and Gordon Burley, *J. Wash. Acad. Sci.*, **41**: 373 (1951).

Let us next consider the stability fields in terms of depths in the earth rather than P and T. To do this we must first establish the average pressure and temperature gradient for the earth. Since the densities of rocks in the earth's crust average very close to 2.8 g/cm³, the pressure rises 280 atm/km.   The mean thermal gradient near the earth's surface averages 20 deg/km.   Since the heat generated by the radioactive elements within the crust accounts for about one-half of that reaching the surface, the gradient at the base of the crust (i.e., at about 30-km depth) should be about one-half the surface value, or 10 deg/km.   These gradients correspond to 14 and 28 atm/deg, respectively.   Thus the thermal gradient in the earth's crust is of a similar magnitude to that of the phase-boundary slopes for a number of the reactions listed in Table 9.1.

In Fig. 9.6 the normal crustal gradient is compared with the phase boundary for calcite and aragonite.   As the locus of temperature and pressure within the crust up to 12 kbars falls within the stability field of calcite, the presence of aragonite in crustal rocks suggests either that the aragonite formed metastably or that it formed in the crust in an area of unusually low thermal gradient (i.e., low temperature).   Most of the aragonite found at the earth's surface was precipitated by marine organisms which for some unknown reason find it more convenient to precipitate the metastable form of $CaCO_3$ rather than calcite.

The presence of the mineral aragonite in the Franciscan metamorphic rocks found in California cannot be due to direct organic precipitation. Detailed studies of these rocks by Ernst (1965) show that aragonite was formed from calcite during the metamorphism of sedimentary rocks. Ernst postulates rapid accumulation of sediment in an oceanic trench

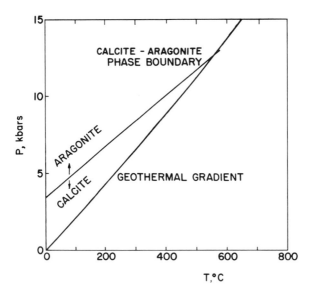

Fig. 9.6. Phase boundary for the transformation calcite-aragonite compared with a normal geothermal gradient.

as the means of producing the high-pressure and low-temperature conditions needed to transform calcite to aragonite.

PERTURBATION OF PHASE BOUNDARIES

The equilibrium position within the earth of any phase transformation depends on the manner in which temperature and pressure increase with depth in the earth. Because of dynamic effects such as convection and because of differences in the amount and vertical distribution of radio-activity, the earth's temperature versus depth curve may be very different from one area to another. Sedimentation, erosion, glaciation, and so forth lead to further temporal changes in both pressure and temperature. Let us consider how glaciation affects the equilibrium depth of a phase boundary.

Neglecting for the moment the heat generated (or released) as the result of a phase change, we can write the following relationship among the change in the depth of the phase boundary, Δx, the earth's pressure gradient, dP/dx, the earth's thermal gradient, dT/dx, the slope of the phase boundary, $(dP/dT)_{\text{P.B.}}$, and the pressure exerted by the glacier, P_G:

$$\Delta x = \frac{P_G}{(dP/dT)_{\text{P.B.}}\, dT/dx - dP/dx}$$

For the earth's upper mantle dP/dx is about 350 atm/km. Pleistocene ice sheets were about 2500 m thick; hence they exerted a pressure of about 250 atm. As stated above, many phase boundaries have a slope of about 18 atm/deg. Thus

$$\Delta x = \frac{250}{18(dT/dx) - 350} \cong \frac{14}{(dT/dx) - 20}$$

Below the base of the crust the thermal gradient should not exceed 15 deg/km; consequently, Δx should lie in the range -0.7 to -3.0 km. Thus, if upon glaciation the phase boundaries in the upper mantle adjust to the increased pressure, they should migrate upward 2 ± 1 km. If the fractional volume change for the solid-solid reaction is 10 percent, then the earth's surface would be lowered 200 ± 100 m.

Since heat is released during the phase transformation, the result obtained above is not correct. The temperature of the rock being transformed will increase by an amount $\Delta H/C_P$. Since the reaction takes place under near-equilibrium conditions, ΔH approximately equals $T\, \Delta S$.

From the Clapeyron equation,

$$\Delta S = \Delta V \left(\frac{dP}{dT}\right)_{\text{P.B.}}$$

Hence complete conversion from the low- to the high-density phase will lead to a temperature rise ΔT given by

$$\Delta T = \frac{T(dP/dT)_{\text{P.B.}}}{C_P/V} \frac{\Delta V}{V}$$

Taking T to be 800°K, $(dP/dT)_{\text{P.B.}}$ to be 18 atm/km, and C_P/V to be 40 atm/deg (since the volume and high-temperature heat capacities of atoms in silicates are similar, this ratio does not vary by more than ± 10 percent), we obtain

$$\Delta T = 360 \frac{\Delta V}{V} \text{ deg}$$

For a reaction where $\Delta V/V$ is 0.10 the temperature will rise by 36 deg.

As shown in Fig. 9.7, such an increase in temperature would cause the phase boundary to be recrossed. What will happen then is that the conversion of low- to high-density material at any depth will proceed until the temperature increase returns the material to the phase boundary. At this point the reaction will cease and the high- and low-density phases will coexist at equilibrium. The zone between the original and final position calculated above will then show a gradation from 100 percent low-density phase at the top (calculated position of new boundary) to some fraction, f, at the base (position of old boundary). The value of f is given by the following equation

$$f = \frac{-\Delta x(dP/dx)(dT/dP)_{\text{P.B.}}}{[T(dP/dT)_{\text{P.B.}} \cdot (\Delta V/V)]/(C_P/V)}$$

where the numerator is the temperature increase required to reach the phase boundary and the denominator the temperature increase for 100 percent conversion of low- to high-density material. Rearranging terms, f becomes

$$f = \frac{(C_F/V) \, \Delta x(dP/dx)}{T(dP/dT)_{\text{P.B.}}^2(\Delta V/V)}$$

Substituting the expression for Δx,

$$f = \frac{-(C_P/V)(dP/dx)P_G}{T(\Delta V/V)(dP/dT)_{\text{P.B.}}{}^2[(dP/dT)_{\text{P.B.}}dT/dx - dP/dx]}$$

For transformation at depths where $dP/dx \gg (dP/dT)_{\text{P.B.}}(dT/dx)$ we have

$$f = \frac{(C_P/V)P_G}{T(dP/dT)_{\text{P.B.}}{}^2(\Delta V/V)}$$

Using the values given above,

$$f = \frac{0.04}{\Delta V/V}$$

Thus for any phase transformation having a large enough $\Delta V/V$ to fall into the 18-atm/deg category, f will be less than unity.   In such cases the lowering of the earth's surface, Δz, is given by

$$\Delta z = \frac{f}{2}\frac{\Delta V}{V}\Delta x$$

Substituting for f we get

$$\Delta z = \frac{(C_P/V)P_G}{2T(dP/dT)_{\text{P.B.}}{}^2}\Delta x$$

Finally, substituting the limiting relationship for Δx, we obtain

$$\Delta z = \frac{(C_P/V)P_G{}^2}{2T(dP/dT)_{\text{P.B.}}{}^2(dP/dx)}$$

Hence for the values used above

$$\Delta z = 14 \text{ m}$$

This result is independent of the magnitude of the volume change.   The corresponding displacement of the phase-boundary position will be

$$\Delta x = \frac{2T(dP/dT)_{\text{P.B.}}{}^2}{P_G(C_P/V)}\Delta z$$

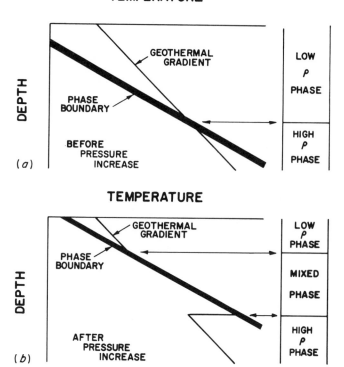

TEMPERATURE

GEOTHERMAL GRADIENT

PHASE BOUNDARY

DEPTH

BEFORE PRESSURE INCREASE

(*a*)

LOW
ρ
PHASE

HIGH
ρ
PHASE

TEMPERATURE

GEOTHERMAL GRADIENT

PHASE BOUNDARY

DEPTH

AFTER PRESSURE INCREASE

(*b*)

LOW
ρ
PHASE

MIXED PHASE

HIGH
ρ
PHASE

Fig. 9.7. *(a)* Earth temperature versus depth and phase-transformation temperature versus depth curves in the region adjacent to a phase boundary prior to glacial loading. *(b)* Same curves after glacial loading. The increased pressure at all depths raises the position of the phase-transformation temperature curve. In the depth range between the final and initial intersections of the two curves the low-density phase is partially transformed into its high-density equivalent. As the transformation proceeds, the earth temperature rises until the temperature at which the two phases coexist at equilibrium is achieved. At this point the reaction ceases, leaving a mixture of the two phases. In the region where the mixed phases exist, the earth temperature and phase-transformation temperatures coincide. The base of the mixed layer corresponds to the initial intersection of the geothermal gradient with the phase boundary.

With the values given above for Δz, P_G, C_P/V, and $(dP/dT)_{\text{P.B.}}$ the displacement in meters is about equal to the absolute temperature in degrees.

RATES

The rates at which solid-solid transformations occur depend not only on the nature of the reaction but also on the ambient temperature and pressure. Davis and Adams (1965) showed, for example, that at 1 atm, 450°C the response time for conversion of aragonite to calcite is of the

TEMPERATURE

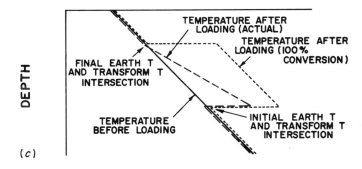

(c)

TEMPERATURE

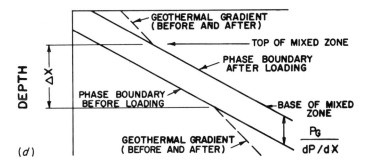

(d)

Fig. 9.7. (c) The initial and final earth-temperature gradients are compared with that which would be produced by the self-heating created by complete transformation of the low- to the high-density phase in the depth region between the initial- and final-temperature curve intersections. (d) Comparison of the initial and final depth versus phase-transformation temperature curves. Those portions of the earth-temperature curve in regions where the phase stability remains unchanged as the result of glacial loading are also shown. Their terminations with the phase-transformation curves define the base and top of the mixed-phase zone (width Δx). The upward displacement of the phase-transformation curve is given by the pressure increase, P_G, divided by the earth pressure gradient, dP/dx.

order of 1 min. They also showed that the activation energy at this pressure is about 100 kcal/mole. Thus the response time at room temperature would be 4×10^{37} years and at 250°C, 20,000 years. It was also shown that by increasing the pressure to 8 kbars the conversion rate was decreased by more than an order of magnitude.

REACTIONS INVOLVING THE RELEASE OF A GAS

Not all reactions taking place between minerals in the earth's crust and mantle are of the simple solid-solid variety we have just described. The materials deposited in the world's oceans and lakes are made up largely of minerals which contain either bound H_2O or bound CO_2. The chief H_2O-bearing minerals are the layered silicates (clays) which make up roughly half of all sediments. Organisms precipitate $CaCO_3$ in almost all natural waters. The calcite and aragonite formed in this way constitute roughly 10 to 20 percent of sedimentary material. Deposits from hypersaline water bodies are rich in gypsum (hydrated $CaSO_4$) and dolomite ($CaMg[CO_3]_2$). As these minerals become buried under the continuing rain of younger debris they are subject to higher pressures and higher temperatures. Increased pressure favors retention of the bound volatiles (they occupy less space when bound). Increased temperature favors their release (the volatilization leads to an entropy increase). As we shall see, for some reactions the normal thermal gradient leads to release of the volatile phase whereas for others an additional heat source is required.

Before considering the complex natural situation, let us perform a hypothetical laboratory experiment. Grains of the minerals quartz (SiO_2) and calcite ($CaCO_3$) are placed in an evacuated and sealed container. The temperature is slowly increased to 1000°K. Either of two reactions might take place:

$$CaCO_3 = CaO + CO_2$$

$$CaCO_3 + SiO_2 = CaSiO_3 + CO_2$$

As both lead to a release of CO_2 gas, the pressure in the container will be a measure of the extent to which one or the other reaction has proceeded.

The free energy of each of the phases can be expressed by

$$G = G^{\text{std}} - \int_{298}^{T} S \, dT + \int_{1}^{P} V \, dP$$

The entropy integral can be written

$$\int_{298}^{T} S \, dT = \int_{298}^{T} S^{\text{std}} \, dT + \int_{298}^{T} \Delta S \, dT$$

where ΔS is the entropy increase which occurs when the substance is heated at constant pressure from 298°K to the temperature T. Thus,

$$\Delta S = \int_{298}^{T} \frac{C_P}{T} \, dT$$

The integral of S^{std} is simply

$$\int_{298}^{T} S^{\text{std}} \, dT = TS^{\text{std}} - 298S^{\text{std}}$$

Substituting into the equation for free energy,

$$G = G^{std} + 298 S^{std} - T S^{std} - \int_{298}^{T} \Delta S \, dT + \int_{1}^{P} V \, dP$$

For the overall reaction,

$$\Delta G = \Delta G^{std} + 298 \, \Delta S^{std} - T \, \Delta S^{std} - \int_{298}^{T} \Delta(\Delta S) \, dT + \int_{1}^{P} \Delta V \, dP$$

The manner in which the free energy difference will change with pressure and temperature can be easily computed if the following simplifying assumptions are made:

1. The volume change for the reaction, ΔV, is approximately equal to the volume of the gas phase, V_{CO_2}.
2. CO_2 behaves as an ideal gas.
3. ΔC_P for the reaction is zero at all temperatures, making ΔS constant with temperature.

If these conditions are met, the free energy change for the reaction will be

$$\Delta G = \Delta H^{std} - T \, \Delta S^{std} + \int_{1}^{p_{CO_2}} V_{CO_2} \, dP$$

If CO_2 behaves as an ideal gas,

$$V_{CO_2} = \frac{RT}{P}$$

and

$$\Delta G = \Delta H^{std} - T \, \Delta S^{std} + RT \ln p_{CO_2}$$

At equilibrium, ΔG will be zero and the partial pressure of CO_2 will be

$$p_{CO_2} = \exp\left(\frac{\Delta S^{std}}{R}\right) \exp\left(-\frac{\Delta H^{std}}{RT}\right)$$

The entropies and enthalpies for the reactants and products of the two reactions are given in Table 9.3. For both reactions the entropy change is positive (reflecting the higher entropy of the gas phase) and the enthalpy change is positive (heat is released when CO_2 becomes bound into the oxide phase). This leads to an increase in CO_2 pressure with rising temperature because the entropy change of the system caused by the gain of heat by the system adds to the entropy change in the system due to the generation of a gas.

TABLE 9.3

Thermodynamic Data for the Possible Reactions Involving CaCO$_3$ and SiO$_2$

	V_{sol}^{std}, cm^3/mole	C_P^{std}	S^{std}	S^{std}/n	H^{std}, cal
CaO	16.67	10.23	9.5	4.7	
CO$_2$	.....	9.02	51.1	17.1	
Σ_{prod}	16.67	19.25	60.6		
CaCO$_3$	39.94	19.57	22.2	4.4	
Σ_{reac}	39.94	19.57	22.2		
Δ	−23.17	−0.32	38.4		42,310
CaSiO$_3$	39.75	20.38	19.6	3.9	
CO$_2$	.....	9.02	51.1	17.1	
Σ_{prod}	39.75	29.40	60.7		
CaCO$_3$	39.94	19.57	22.2	4.4	
SiO$_2$	22.69	10.62	10.1	3.4	
Σ_{reac}	62.63	30.19	32.3		
Δ	−22.88	−0.79	38.4		21,060
CaSiO$_3$	39.75	20.38	19.6	3.9	
Σ_{prod}	39.75	20.38	19.6		
CaO	16.67	10.23	9.5	4.7	
SiO$_2$	27.69	10.62	10.1	3.4	
Σ_{reac}	39.36	20.85	19.6		
Δ	0.39	−0.47	0.0		−21,250

For the first reaction,

$$p_{CO_2} = e^{19.2} \exp\left(-\frac{21,150}{T}\right)$$

and the second reaction

$$p_{CO_2} = e^{19.2} \exp\left(-\frac{11,530}{T}\right)$$

The predicted pressures are shown as a function of temperature in Table 9.4. At all temperatures the partial pressure of CO$_2$ due to CaCO$_3$ decomposition to CaO is much less than that due to the reaction involving SiO$_2$. Hence in the presence of quartz CaO is always an unstable phase. In other words, the free energy of CaSiO$_3$ is always lower than the free energy of CaO + SiO$_2$. At standard conditions the free energy change for the reaction CaO + SiO$_2$ → CaSiO$_3$ is −21,150 cal/mole.

TABLE 9.4

Equilibrium Pressure of CO_2 for Reaction of $CaCO_3$ with SiO_2 and for Decomposition of $CaCO_3$ on the Assumption that CO_2 Is an Ideal Gas, $\Delta C_P = 0$, and $\Delta V_{sol} = 0$

T, °K	T, °C	$CaCO_3 = CaO + CO_2$ p_{CO_2}, atm	$CaCO_3 + SiO_2 = CaSiO_3 + CO_2$ p_{CO_2}, atm
300	25	6.6×10^{-23}	1.3×10^{-7}
400	125	2.7×10^{-15}	8.2×10^{-4}
500	225	1.0×10^{-10}	0.17
600	325	1.0×10^{-7}	5.4
700	425	1.6×10^{-5}	67
800	525	8.2×10^{-4}	4.4×10^2
900	625	1.3×10^{-2}	1.8×10^3
1000	725	0.15	6.0×10^3

Were we to examine the contents of the vessel we would find that for each mole of CO_2 gas present there would be 1 mole of wollastonite. Hence for any given pressure the amount of the products would depend on the size of the chamber. Also, for each mole of CO_2 manufactured, 1 mole of quartz and 1 mole of calcite would disappear. If we had placed a greater number of moles of quartz than of calcite in the chamber we would note that at some point the pressure rise would abruptly depart from the predicted path. This would mark the disappearance of the last bit of calcite. Further increases in temperature would not change the amount of CO_2; however, the pressure would rise in accordance with the equation of state of CO_2. The residual quartz would coexist stably with the wollastonite.

On the other hand, had we placed too little quartz in the chamber we would again see an abrupt departure in pressure rise at the point at which the quartz was entirely consumed. Beyond the temperature at which this occurred calcite and wollastonite would stably coexist. However, a continuation of the temperature increase would eventually lead to decomposition of the $CaCO_3$ to CaO when the equilibrium pressure for this reaction became equal to the pressure of CO_2 in the chamber. The CO_2 pressure would again begin to rise steeply along the $CaCO_3$ decomposition curve until the calcite was consumed. Once this had occurred the amount of CO_2 would again become constant and CaO and $CaSiO_3$ would stably coexist. This sequence is shown graphically in Fig. 9.8.

Were we to install a pressure release valve in the chamber (say at 500 atm) the sequence of events would be somewhat different. The CO_2 pressure would rise as before until it reached 500 atm if enough calcite were available. At this point the CO_2 would vent and the reaction would proceed until either the quartz or the calcite was entirely consumed. If the calcite were depleted first, quartz and wollastonite would stably coexist as the temperature continued to rise. Were quartz first to be depleted, wollastonite and calcite would coexist with rising temperature until the point at which the decomposition pressure of calcite reached 500 atm. At this point more CO_2 would be rapidly vented and the $CaCO_3$ would give way to CaO. Beyond this temperature CaO and $CaSiO_3$ would coexist (on the assumption that $CaO + CaSiO_3$ is a more stable combination than, for example, Ca_2SiO_4).

Thus we see that the course of the reaction depends upon (1) the ratio of the reactants, (2) the ability of the chamber to hold the CO_2, and (3) the ratio of the volume available to the gas to that occupied by

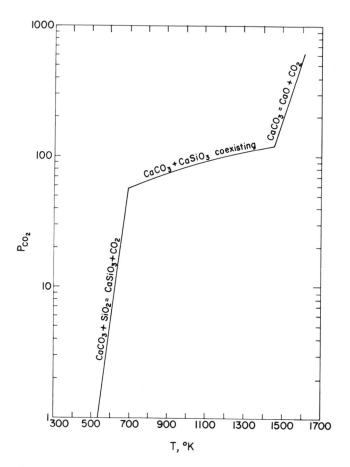

Fig. 9.8. The change in CO_2 pressure during the reaction of $CaCO_3$ to form $CaSiO_3$ and CaO. See text for discussion.

the solids.   We can, however, make some generalizations regarding the equilibrium state.   (1) If the partial pressure of CO_2 is equal to that predicted for the reaction of SiO_2 + $CaCO_3$ to produce $CaSiO_3$, then all three solid phases must be present in the chamber.   (2) If the p_{CO_2} is lower than the predicted value, either quartz or calcite must be absent.   (3) If the CO_2 pressure exceeds the equilibrium pressure, as it would if we had initially pumped the chamber full of CO_2 gas, then wollastonite will be absent.   Above or below the locus of equilibrium p_{CO_2} $-$ T points, only three phases can coexist (two solids and CO_2 gas).   On the equilibrium curve there can be four (three solids and CO_2 gas).

Having seen the general character of reactions involving the release of volatiles, let us go back and consider the differences which will take place if we treat the system rigorously.   The major differences are that CO_2 is not an ideal gas, the heat-capacity difference is not zero (see Table 9.3), and at sufficiently high pressures the volume change for the solids becomes a significant factor.

Let us first consider the heat-capacity effects.   As shown in Table 9.5, the difference in heat capacity between the reactants and products increases with rising temperature up to the low- to high-quartz transition at $848°K$.   This increase reflects the high vibrational frequencies in CO_2 gas which lead to a much lower fractional increase in its heat capacity relative to that for the three solids.   Therefore, the entropies of the reactants (quartz + calcite) rise more rapidly than those of the products (CO_2 + wollastonite).

TABLE 9.5

Heat-capacity Data for Reaction of $CaCO_3$ with SiO_2[†]

Temp, °K	C_P CO_2	C_P $CaSiO_3$	C_P Σ_{prod}	C_P $CaCO_3$	C_P SiO_2	C_P Σ_{reac}	ΔC_P
300	9.0	20.4	29.4	19.6	10.6	30.2	−0.8
400	10.1	24.0	34.1	23.2	12.9	36.1	−2.0
500	10.8	25.8	36.6	25.1	14.3	39.4	−2.8
600	11.3	27.0	38.3	26.4	15.4	41.9	−3.6
700	11.6	27.8	39.4	27.4	16.4	43.8	−4.4
800	11.9	28.5	40.4	28.2	17.4	45.6	−5.2
900	12.2	29.1	41.3	28.9	16.2	45.1	−3.8
1000	12.5	29.6	42.1	29.6	16.4	46.0	−3.9

† Data taken from A. Danielsson, *Geochim. Cosmochim. Acta*, **1**: 55–69 (1950).

$CaSiO_3$	$26.64 + (3.60 \times 10^{-3}T) - (6.5 \times 10^5 T^{-2})$
CO_2	$10.55 + (2.16 \times 10^{-3}T) - (2.04 \times 10^5 T^{-2})$
$CaCO_3$	$24.98 + (5.24 \times 10^{-3}T) - (6.20 \times 10^5 T^{-2})$
Low SiO_2 ($<848°K$)	$11.22 + (8.20 \times 10^{-3}T) - (2.70 \times 10^5 T^{-2})$
High SiO_2 ($>848°K$)	$14.41 + (1.94 \times 10^{-3}T)$

The 38.4-cal/deg-mole entropy difference at 300°K is reduced by 4.0 cal/deg-mole to 34.4 cal/deg-mole at 1000°K (see Table 9.6). As stated above, this change is given by the integral

$$\Delta(\Delta S) = \int_{298}^{T} \frac{\Delta C_P}{T} \, dT$$

The corrected free energy equation becomes

$$\Delta G = \Delta H^{\text{std}} - T \, \Delta S^{\text{std}} - \int_{298}^{T} \Delta(\Delta S) \, dT + RT \ln p_{\text{CO}_2}$$

The equilibrium p'_{CO_2} pressure becomes

$$p'_{\text{CO}_2} = \exp\left(\frac{\Delta S^{\text{std}}}{R}\right) \exp\left(-\frac{\Delta H^{\text{std}}}{RT}\right) \exp\left(\frac{\int_{298}^{T} \Delta(\Delta S) \, dT}{RT}\right)$$

Designating the partial pressure of CO_2 calculated in previous idealized calculations as p_{CO_2}, we have

$$p'_{\text{CO}_2} = p_{\text{CO}_2} \exp\left(\frac{\int_{298}^{T} \Delta(\Delta S) \, dT}{RT}\right)$$

Values of the correction term are given in Table 9.6 and the corrected pressures in Table 9.7. At 1000°K the corrected pressure is one-half that obtained by assuming the entropy difference for the reaction to remain constant with temperature.

Before we go on to the volume effects, one other thermal correction must be made. At 848°K low quartz transforms to high quartz. For each mole of quartz, 290 cal must be added to the system to accomplish this change. The entropy change, ΔS, for the reaction is reduced by $\frac{290}{848}$ or 0.34 cal/deg-mole. This correction has been included in Table 9.6.

Surprisingly enough, the ideal gas assumption is nearly valid to 900°K for CO_2. The ratio of the actual gas volume to that obtained from the ideal gas law is shown in Table 9.8. Up to 700 deg the equilibrium CO_2 pressure is sufficiently low so that the gas is indeed ideal. In the range 700 to 1000°K CO_2 has a temperature close to its Boyle temperature (1025°K). As stated in Chap. 2, under these conditions the two major sources of nonideality nearly cancel. Beyond 900°K, however, the finite volume of the CO_2 molecules begins to dominate the attraction effect, and the actual volume becomes significantly greater than that given by the ideal gas law. Beyond this temperature the equilibrium

TABLE 9.6
Entropy Contribution to the Reaction of CaCO$_3$ with SiO$_2$

Temp, °K	ΔC_P, cal/deg·mole	$\Delta C_P/T \times 10^3$, cal/deg^2·mole	$\Delta(\Delta S)$, cal/deg·mole	ΔS, cal/deg·mole	$\int\Delta(\Delta S)\,dT$, cal/mole	$\dfrac{\int\Delta(\Delta S)\,dT}{RT}$	$\exp\left(\dfrac{\int\Delta(\Delta S)\,dT}{RT}\right)$
300	−0.8	−2.7	−0.0	38.4	−0	−0.00	1.00
400	−2.0	−5.0	−0.4	38.0	−20	−0.03	0.97
500	−2.8	−5.6	−0.9	37.5	−90	−0.09	0.91
600	−3.6	−6.0	−1.5	36.9	−220	−0.18	0.84
700	−4.4	−6.3	−2.1	36.3	−400	−0.29	0.75
800	−5.2	−6.5	−2.7	35.7	−640	−0.40	0.67
			Low- to High-quartz Inversion				
900	−3.8	−4.3	−3.6†	34.8†	−960	−0.53	0.60
1000	−3.9	−3.9	−4.0†	34.4†	−1350	−0.67	0.50

† Includes entropy change due to inversion.

TABLE 9.7

Equilibrium Pressure of CO_2 for Reaction of $CaCO_3$ with SiO_2 as in Table 9.4 Compared with Equilibrium Pressure Including the Entropy Correction

Temp, °K	p_{CO_2}	p'_{CO_2}
300	1.3×10^{-7}	1.3×10^{-7}
400	8.2×10^{-4}	8.0×10^{-4}
500	0.17	0.15
600	5.4	4.5
700	67	50
800	440	300
900	1800	1070
1000	6000	3000

pressure will rise much more slowly than predicted by our idealized equation.

Before attempting to correct for this effect, let us consider the importance of the volume change for the solid phases. As shown in Table 9.3, wollastonite occupies 2.3×10^{-2} liter/mole less volume than do quartz + calcite (that is, $\Delta V^{sol} = -2.3 \times 10^{-2}$ liter/mole). If the pressure exerted on the solids is only that due to CO_2, the correction becomes

$$\Delta(\Delta G) = \Delta V^{sol} p_{CO_2}$$

TABLE 9.8

Reduced Temperature and Pressure for CO_2 Produced in the Reaction of $CaCO_3$ with SiO_2 and the Ratios of Actual Volume to Volume Obtained by Assuming CO_2 Is Ideal†

Temp, °K	p_{CO_2}, atm	T_R	P_R	V_{actual}/V_{ideal}
300	1.3×10^{-7}	1.0	1.8×10^{-9}	1.00
400	8.0×10^{-4}	1.3	1.1×10^{-5}	1.00
500	0.15	1.6	2.1×10^{-3}	1.00
600	4.5	2.0	0.062	1.00
700	50	2.3	0.69	0.99
800	300	2.6	4.1	1.03
900	920	3.0	13	1.25
1000	2700	3.3	37	1.95

† Data taken from A. Danielsson, *Geochim. Cosmochim. Acta*, **1**: 55–69 (1950).

When this term is incorporated into the CO_2 equation, we obtain

$$p''_{CO_2} = \exp\left(\frac{\Delta S^{std}}{R}\right) \exp\left(-\frac{\Delta H^{std}}{RT}\right) \exp\left(\frac{\int \Delta(\Delta S)\, dT}{RT}\right) \exp\left(-\frac{\Delta V^{std} p''_{CO_2}}{RT}\right)$$

Using p'_{CO_2} as the approximate pressure,

$$p''_{CO_2} = p'_{CO_2} \exp\left(-\frac{\Delta V^{std} p'_{CO_2}}{RT}\right)$$

As shown in Table 9.9, the correction becomes important only above $800°K$ (300 atm). It leads to an equilibrium CO_2 pressure higher than that given by the simple model.

TABLE 9.9

**Effect of Volume Change of Solids on the
Reaction of $CaCO_3$ with SiO_2**

Temp, °K	ΔV^{sol}, 10^{-2} liter/mole	$\Delta V^{sol}/RT$, 10^{-4} atm^{-1}	$\Delta V^{sol} p_{CO_2}/RT$	$\exp\left(-\dfrac{\Delta V^{sol}\, p'_{CO_2}}{RT}\right)$
300	−2.3	−9.2	$−1.2 \times 10^{-10}$	1.00
400	−2.3	−6.9	$−6.4 \times 10^{-7}$	1.00
500	−2.3	−5.5	$−8.2 \times 10^{-5}$	1.00
600	−2.3	−4.6	$−2.1 \times 10^{-3}$	1.00
700	−2.3	−3.9	$−2.0 \times 10^{-2}$	1.02
800	−2.3	−3.4	−0.10	1.10
900	−2.3	−3.1	−0.29	1.34

It is thus clear that for temperatures above $800°K$ the use of the ideal gas law and neglect of the volume change of the solids will lead to serious errors in the estimate of equilibrium CO_2 pressures. In the range 800 to $1000°K$ the volume of CO_2 gas can be approximated as follows:

$$V_{CO_2} = \frac{RT}{p_{CO_2}} + b$$

where b is about 2.8×10^{-2} liter/mole. Using this relationship, the free energy equation becomes

$$\Delta G = \Delta H^{std} - T\,\Delta S^{std} - \int_{298}^{T} \Delta(\Delta S)\, dT + RT \ln p_{CO_2} + (\Delta V^{std} + b) p_{CO_2}$$

At equilibrium

$$p_{CO_2} = \exp\left(\frac{\Delta S^{std}}{R}\right) \exp\left(-\frac{\Delta H^{std}}{RT}\right) \exp\left(\frac{\int_{298}^{T} \Delta(\Delta S)\, dT}{RT}\right)$$
$$\exp\left(-\frac{\Delta V^{std} + b}{RT}\, p_{CO_2}\right)$$

In terms of p'_{CO_2},

$$p''_{CO_2} = p'_{CO_2} \exp\left(-\frac{\Delta V^{std} + b}{RT}\, p''_{CO_2}\right)$$

Values for p''_{CO_2} can be obtained by successive approximation (see Table 9.10). As the finite volume of the CO_2 molecules (28 cm³/mole) nearly balances the volume change of the solids (23 cm³/mole) the net correction is small. Thus the result obtained using the ideal gas law and neglecting the volume change of the solids is not so bad after all. As long as the volume of the gas molecules approximates the net volume change of the solids, this approach is valid. In the example chosen here it works all the way up to 3000 atm.

TABLE 9.10

Equilibrium Pressure of CO₂ for Reaction of CaCO₃ with SiO₂ Including Effects of Entropy, Nonideality of CO₂, and Volume Change by Solids

Temp, °K	$\Delta V^{sol} + b$, 10⁻² liter/mole	$\dfrac{\Delta V^{sol} + b}{RT}$, 10⁻⁴ atm⁻¹	p'_{CO_2}	$\dfrac{\Delta V^{sol} + b}{RT} p'_{CO_2}$	$\exp\left(-\dfrac{\Delta V^{sol} + b}{RT} p'_{CO_2}\right)$
800	0.5	0.77	300	0.02	0.98
900	0.5	0.68	1070	0.07	0.93
1000	0.5	0.61	3000	0.18	0.83

Temp, °K	First Estimate, p''_{CO_2}	$\exp\left(-\dfrac{\Delta V^{sol}}{RT} p''_{CO_2}\right)$	Second Estimate, p''_{CO_2}
800	293	0.98	293
900	1000	0.93	1000
1000	2500	0.85	2600

Having established the equilibrium pressure of CO_2 gas over a mixture of quartz and calcite as a function of temperature, we turn to the question of where in the earth's crust we might expect quartz and calcite to combine to form wollastonite. We might compare any unit of rock to a pressure cooker.   Below some critical pressure it will act as a closed system, retaining any CO_2 produced by this reaction.   Since the free volume (porosity) of earth rocks is extremely small, the CO_2 required to generate the equilibrium pressure will not measurably deplete the supply of quartz and calcite.   For all practical purposes, we can say that until a critical temperature is achieved the reaction does not occur.   When the equilibrium CO_2 pressure exceeds the venting pressure of the rocks, CO_2 will escape and the reaction will proceed until either the quartz or calcite is entirely consumed.   In this sense the reaction is an "either-or" proposition; only when the equilibrium partial pressure of CO_2 exceeds the confining pressure of the rock will the reaction proceed to a significant extent.

The CO_2 pressure required to cause venting depends not only on the depth to which a rock is buried but also on other more subtle variables. Two extreme situations will demonstrate why this is the case.   First, consider a situation where the rock of interest is buried under 1000 m of unlithified sand.   The pores in the sand are water-saturated.   In order for CO_2 generated at the base of the sand to escape, it would have to be able to lift a column of water 1000 m high.   Bubbles could then form and migrate to the surface.   If the density of the water is assumed to be uniformly 1 g/cm^3, a pressure of 100 atm would be necessary.

Next consider a water-filled void completely enclosed in a compact siliceous limestone.   In order for the CO_2 to escape, the pressure would have to rise to the point where the vesicle would rupture.   This could certainly not occur until the pressure reached the level where it balanced the weight of the overlying rock.   Until this point the vesicle would be under compressional stress.   As the rock would have some finite tensile strength, additional pressure would be required to induce rupture.   Even if this overpressure were considered small, the release pressure would be roughly 2.5 times (the ratio of rock to water density) as high as in the case of the water-saturated sand.

Most natural systems probably lie somewhere between these extremes.   More pressure is needed than that required to lift a column of water equal in length to the depth of burial and less than is necessary to lift a column of rock.   Other factors to be considered are the friction to the flow of the discharged fluids and the contribution of water vapor to the total pressure of the gas phase.   The former leads to higher critical CO_2 pressures and the latter to lower pressures.

Situations could certainly arise where the pressure exerted on the solid phases exceeded that exerted on the included gases.   The sand grains at the base of the hypothetical column discussed above would be supporting the weight of the overlying sand whereas adjacent water molecules would be supporting the much lower weight of the overlying water column.   For such cases this difference should be taken into account in the free energy calculations. The volume change for the solids should be multiplied by the lithostatic pressure rather than the ambient CO_2 pressure. Hence the free energy equation would be revised to read

$$\Delta G = \Delta H^{std} - T \, \Delta S^{std} - \int_{298}^{T} \Delta(\Delta S) \, dT + \Delta V^{std} P_{lith} + b p_{CO_2} + RT \ln p_{CO_2}$$

We are now in a position to estimate under what conditions in the earth's crust calcite and quartz would react to yield wollastonite. In Fig. 9.9 a plot of a typical geothermal gradient for the earth's crust is compared with curves for the minimum temperature (release pressure equals weight of equivalent water column) and maximum temperature (release pressure equals weight of equivalent rock column) required for wollastonite production. Clearly this reaction does not take place as the result of normal geothermal heating. Abnormally high temperatures as would be generated, for example, adjacent to igneous intrusions would be required.

In Table 9.11 the thermodynamic data for several CO_2 release reactions are compared. Despite the fact that the decomposition temperatures show an extremely large range (that is, 1000°K), the standard entropy, volume, and heat-capacity changes are all fairly similar. The

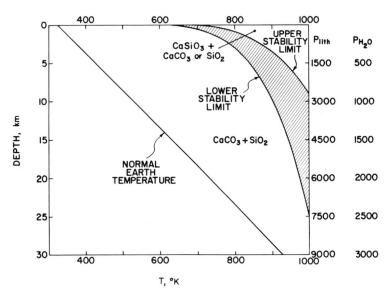

Fig. 9.9. Location in the earth's crust where the reaction calcite + quartz → wollastonite might occur.

TABLE 9.11

Thermodynamic Data and Decomposition Temperature for Several Solids Which Decompose to Release CO_2†

Reaction	ΔH^{std}, kcal/mole	ΔS^{std}, cal/deg-mole	ΔC_P^{std}, cal/deg-mole	ΔV^{sol}, cm^3	Decompos. Temp, 1 atm, °K	Decompos. Temp, 300 atm, °K
$ZnCO_3 = ZnO + CO_2$	16.9	41.8	−0.6	−13.9	405	550
$PbCO_3 = PbO + CO_2$	20.9	35.4	−1.1	−16.7	590	860
$SiO_2 + CaCO_3 = CaSiO_3 + CO_2$	21.2	38.5	−0.9	−19.6	550	775
$MgCO_3 = MgO + CO_2$	28.1	42.0	−0.1	−16.8	670	910
$CaCO_3 = CaO + CO_2$	42.5	38.4	−0.5	−20.2	1100	1560
$SrCO_3 = SrO + CO_2$	56.1	40.9	+0.2	−17.0	1370	1890
$BaCO_3 = BaO + CO_2$	63.9	41.1	−0.2	−19.0	1550	2140

† Data for entropy and heat capacity from K. K. Kelley and E. G. King, *U.S. Bur. Mines Bull.* 592, 1961. Volume data except for BaO and SrO from S. P. Clark, Jr. (ed.), Handbook of Physical Constants, *Geol. Soc. Am. Mem.* 97, 1966. Volumes for SrO and BaO from "Handbook of Chemistry and Physics," The Chemical Rubber Publishing Company, Cleveland, Ohio. Enthalpy data except for ZnO from R. M. Garrels and C. L. Christ, "Solutions, Minerals, and Equilibria," Harper & Row, Publishers, Incorporated, New York, 1965; for ZnO from Handbook of Physical Constants, *Geol. Soc. Am. Mem.* 97, 1966.

big difference in their tendencies toward decomposition arises from the variation in ΔH^{std} from reaction to reaction.

The question comes to mind whether the tendency of a given carbonate to decompose can be predicted from its properties and those of the oxide produced by its breakdown. In Table 9.11 it is shown those carbonate-oxide transformations showing the largest decrease in heat capacity decompose most easily. The magnitude of the heat-capacity decrease is related to the relative strengths of the metal oxide bond in the carbonate versus that in the oxide. A large decrease reflects enhanced bond strength in the oxide relative to the carbonate. Such an enhanced bond strength should lead to a larger oxide stability field and a lower decomposition temperature.

It is of interest to note that the decomposition temperature rise is in accordance with the degree of covalency of the metal oxygen bonds. A high degree of covalency (Zn, for example) leads to easy decomposition, whereas a high degree of ionic character (Ba, for example) leads to difficult decomposition.

PROBLEMS

9.1 A phase transition occurs at $P = 4$ kbars, $T = 0°C$. The entropy change for the transition is -2 cal/deg-mole, and the volume decrease is 6 cm³/mole. Assuming a normal geothermal gradient, at what pressure in the earth will the transition occur?

*9.2 Oldhamite, CaS, is found only in enstatite-rich meteorites. Larimer has studied the reaction

$$\tfrac{1}{2}CaS + O_2 = \tfrac{1}{2}CaSO_4$$

and found that log $p_{O_2} = 8.58 - 25{,}350/T°K$ for T between 800 and 1000°C. Using free energy and entropy data, calculate the theoretical O_2 pressure as a function of temperature. Compare theoretical and experimental results.

*9.3 The Moho under oceanic regions occurs at a depth of about 12 km below sea level. If the seismic discontinuity were due to a phase change, at what depth below the surface of the moon would a similar phase change occur? If identical chemical composition is assumed, would there be any difference in the thermal effects of the transition in the moon as compared with those in the earth?

9.4 A detailed study of a large alkali feldspar crystal reveals a fluid inclusion. The feldspar consists of two phases, one rich in K (as $KAlSi_3O_8$), the other rich in Na (as $NaAlSi_3O_8$). The fluid inclusion contains H_2O, K^+, and Na^+.

  (a) What is the number of components necessary to define the system?

  (b) How many phases are present?

  (c) How many independent equations can be written connecting the components and the phases?

  (d) If the pressure and composition of all the phases are given, is the system completely defined? If not, what additional information is necessary?

  (e) Would simple application of the phase rule, in the form $f = C + 2 - P$, be likely to illuminate or confuse the essential nature of the problem?

REFERENCES

Boyd, F. R., and J. L. England: *J. Geophys. Res.*, **65**: 749 (1960).

Danielsson, A.: *Geochim. Cosmochim. Acta*, **1**: 55 (1950).

Davis, B. L., and L. H. Adams: *J. Geophys. Res.*, **70**: 433 (1965).

Ernst, W. G.: *Geol. Soc. Am. Bull.*, **76**: 879 (1965).

Kracek, F. C., N. J. Neuvonen, and Gordon Burley: *J. Wash. Acad. Sci.*, **41**: 373 (1951).

Waldbaum, David R.: *Am. Mineralogist*, **50**: 186 (1965).

SUPPLEMENTARY READING

Ahrens, T. J., and Y. Syono: Calculated Mineral Reactions in the Earth's Mantle, *J. Geophys. Res.*, **72**: 4181 (1967).

Boettcher, A. L., and P. J. Wyllie: Jadeite Stability Measured in the Presence of Silicate Liquids in the System $NaAlSiO_4$-SiO_2-H_2O, *Geochim. Cosmochim. Acta*, **32**: 999 (1968).

Boyd, F. R.: Petrologic Problems in High-pressure Research, in P. H. Abelson (ed.), "Researches in Geochemistry," vol. 2, p. 593, John Wiley & Sons, Inc., New York, 1967.

Eugster, H. P., and G. B. Skippen: Igneous and Metamorphic Reactions Involving Gas Equilibria, in P. H. Abelson (ed.), "Researches in Geochemistry," vol. 2, p. 492, John Wiley & Sons, Inc., New York, 1967.

Goldsmith, J. R.: Metastability and Hangovers in Crystals, *Geochim. Cosmochim. Acta*, **31**: 913 (1967).

Green, D. H., and A. E. Ringwood: An Experimental Investigation of the Gabbro to Eclogite Transformation and Its Petrological Applications, *Geochim. Cosmochim. Acta*, **31**: 767 (1967).

Kushiro, I., Y. Syono, and S. Akimoto: Stability of Phlogopite at High Pressures and Possible Presence of Phlogopite in the Earth's Upper Mantle, *Earth Planet. Sci. Letters*, **3**: 197 (1967).

Larimer, J. W.: An Experimental Investigation of Oldhamite, CaS; and the Petrologic Significance of Oldhamite in Meteorites, *Geochim. Cosmochim. Acta,* **32:** 965 (1968).

Newton, R. C.: Kyanite-Sillimanite Equilibrium at 750°C, *Science,* **151:** 1222 (1966).

———: Kyanite-Andalusite Equilibrium from 700° to 800°C, *Science,* **153:** 170 (1966).

*O'Connell, R. J., and G. J. Wasserburg: Dynamics of the Motion of a Phase Change Boundary to Changes in Pressure, *Rev. Geophys.,* **5:** 329 (1967).

Ringwood, A. E., and D. H. Green (eds.): Phase Transformations and the Earth's Interior, proceedings of a symposium held in Canberra, Australia, 6–10 January, 1969, North-Holland Publishing Co., Amsterdam, 1970.

Schuiling, R. D., and B. W. Vink: Stability Relations of Some Titanium Minerals (Sphene, Perovskite, Rutile and Anatase), *Geochim. Cosmochim. Acta,* **31:** 2399 (1967).

Wones, D. R.: A Low Pressure Investigation of the Stability of Phlogopite, *Geochim. Cosmochim. Acta,* **31:** 2248 (1967).

chapter ten Melting Phenomena

Evidence of present and past igneous activity testifies to the fact that silicate material in the earth's mantle, and possibly in the lower crust, is continuously being subjected to partial melting. Since the crust and mantle both transmit shear waves which would not travel in fluids, we know that most of the crust and mantle is solid. To understand the process of magma formation we would like to know under what conditions the solid phases present will melt.

The melting point of any given mineral depends in a rather complex way on its environment. Not only does the fusion temperature depend on the pressure to which the mineral is subjected but also on the minerals and fluids with which it is associated. In this chapter we will deal with the factors which influence the melting of solids.

PRESSURE DEPENDENCE OF MELTING POINTS

The melting points of most materials of interest in earth science rise with increasing confining pressure. Ice is a notable exception. Since the thermodynamic properties of silicate materials at high temperature and pressures are not well known, information concerning the behavior of melting points with changing pressure must be derived from experimental data. Observed melting points for several geologically important materials are given as a function of pressure in Table 10.1. The gradients average about 10 deg/kbar. However, it should be noted that the gradients are decidedly nonlinear at high pressures.

TABLE 10.1

Melting Temperature as a Function of Pressure for Several Solids

Mineral	Temperature, °C				Average Gradient, deg/kbar 0–10 kbars
	$P = 0$ kbar	$P = 10$ kbars	$P = 20$ kbars	$P = 30$ kbars	
Diopside ($CaMgSi_2O_6$) †	1390	1520	1630	1710	13
Albite ($NaAlSi_3O_8$) †	1120	1240	1320	1400	12
Enstatite ($MgSiO_3$) ‡	1557 §	1670	1760	1840	(11) §
Forsterite ¶	1900	1950	1990	2040	5

† F. R. Boyd and J. L. England, *J. Geophys. Res.*, **68**: 311 (1963).
‡ F. R. Boyd, J. L. England, and B. T. C. Davis, *J. Geophys. Res.*, **69**: 2101 (1964).
§ Incongruent melting of clinoenstatite; thus, gradient is average of two curves with different slopes.
¶ B. T. C. Davis and J. L. England, *J. Geophys. Res.*, **69**: 1113 (1964).

From the Clapeyron equation, discussed in Chap. 9, we know that the melting-point gradient is related to the entropy of fusion, ΔS_f, and volume change on fusion, ΔV_f:

$$\frac{dT_m}{dP} = \frac{\Delta V_f}{\Delta S_f}$$

Thus the approximate entropy of fusion may be calculated from the measured gradients and the volumes of the solid and melt for the small number of cases where the molar volume of the melt is known. The percentage entropy change for melting averages about 5 times larger than for solid-solid transformations. Since the volume changes occurring during the two types of transitions are about the same, the melting-point gradient (dT_m/dP) must be about 5 times less steep than the slope of the phase boundary for comparable solid-solid transitions.

The slope of any melting curve will change if it is intersected by a solid-solid-transition phase boundary. This can be seen by referring to Fig. 9.1, the phase diagram for SiO_2. The melting-point gradient, dT_m/dP, is much higher for β quartz, the high-density phase, than it is for cristobalite, the low-density phase. From the positive slope of the cristobalite–β-quartz phase boundary we can conclude that the entropy of cristobalite is higher than that of β quartz. Since the entropy of SiO_2 liquid is higher than that of either solid, we can conclude that the entropy of fusion for β quartz is higher than that for cristobalite. The volume change on fusion of β quartz is also larger. The reason that the melting-point gradient increases on passing from cristobalite to β quartz is that the percentage entropy difference between the two solids, as compared with the difference between either solid and the liquid, is less than the percentage volume difference.

To see this point more clearly, let us assume we are dealing with the β-quartz–cristobalite–liquid SiO_2 triple point. We may then express the volume and entropy changes on fusion of β quartz as the sum of the changes in going from β quartz to cristobalite and thence to the liquid:

$$\left(\frac{dT_m}{dP}\right)_{\beta\,\text{qtz}} = \frac{\Delta V_{\beta\,\text{qtz–liq}}}{\Delta S_{\beta\,\text{qtz–liq}}} = \frac{\Delta V_{\text{crist–liq}} + \Delta V_{\beta\,\text{qtz–crist}}}{\Delta S_{\text{crist–liq}} + \Delta S_{\beta\,\text{qtz–crist}}}$$

For melting of cristobalite

$$\left(\frac{dT_m}{dP}\right)_{\text{crist}} = \frac{\Delta V_{\text{crist–liq}}}{\Delta S_{\text{crist–liq}}}$$

The terms $\Delta V_{\text{crist–liq}}$ and $\Delta V_{\beta\,\text{qtz–crist}}$ are approximately the same; however, the term $\Delta S_{\text{crist–liq}}$ is much larger than $\Delta S_{\beta\,\text{qtz–crist}}$. Thus,

$$\left(\frac{dT_m}{dP}\right)_{\beta\,\text{qtz}} > \left(\frac{dT_m}{dP}\right)_{\text{crist}}$$

Because of this phenomenon, care must be taken not to extrapolate experimental data on fusion curves beyond the intersection of the fusion curve with any solid-solid-transition phase boundary.

For a liquid to form in the mantle, the earth's pressure-temperature curve (the geothermal gradient) must cross the composite melting curve for the solid phases present. As will be shown later in this chapter, the presence of more than one solid phase may lower the temperature at which a liquid first appears to a point below the melting point of the pure solids. Nevertheless, it is interesting to compare the melting curve of a mineral which might be found as a major phase in the mantle with the geothermal gradient. Figure 10.1 shows the pressure dependence of the melting point of diopside, a pyroxene, as measured by Boyd and England, together with a geothermal gradient calculated by Clark and Ringwood from consideration of measured terrestrial heat flow. If diopside were

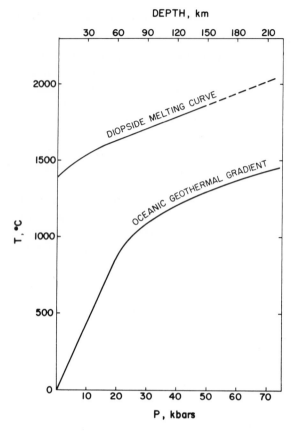

Fig. 10.1. Melting curve for diopside determined by Boyd and England, and a geothermal gradient calculated by Clark and Ringwood from consideration of heat-flow data.

the only solid present, and if this geothermal gradient held for the whole earth, no liquid phase could ever form.

MELTING IN IDEAL MIXED–OXIDE SYSTEMS

When two solid phases are mixed and heated, a liquid will commonly appear at a lower temperature than if either oxide were heated separately. A practical example of this phenomenon is provided by the common use of solid salt to melt the ice on a sidewalk. To understand why this happens, consider an ice cube floating in distilled water at 0°C. As long as no heat is added to or extracted from the container both phases will remain. This does not mean that no interaction between the two occurs, but rather that for every H_2O molecule which jumps free of the ice crystal another one is captured. Now, if we were to allow some table salt to dissolve in the water, we would upset this equilibrium. The dilution of the H_2O molecules in the liquid by Na^+ and Cl^- ions would cause fewer H_2O molecules to hit the ice crystal and hence fewer to be captured. The result would be that the ice would lose H_2O molecules faster than it gained them, and the ice cube would begin to shrink. However, as the melting proceeded, cooling would ensue [to remove H_2O molecules from ice requires $1436/(6 \times 10^{23})$ cal/molecule]. As the ice and salt solution cooled, both the rate of escape and capture would fall. However, the rate of escape would be most affected. Eventually a temperature would be reached where the two rates would again become equal. This melting-point lowering would occur provided two conditions were met: (1) that the melt of the two solids formed a single homogeneous liquid (i.e., liquids miscible) and (2) that the solid contaminant in the liquid phase was not able to dissolve in the solid of interest (i.e., solids immiscible).

These conditions are met for a wide variety of silicate-mineral pairs. In such cases we can calculate the melting-point depression caused by the contaminant in the following way. The free energies of the two solid phases A and B can be written as follows, provided that their entropies are essentially constant near the melting point:

$$G_{A_{sol}} = G^*_{A_{sol}} - (T - T_A)S^*_{A_{sol}}$$

and

$$G_{B_{sol}} = G^*_{B_{sol}} - (T - T_B)S^*_{B_{sol}}$$

where T_A and T_B are the melting points of the pure solids, and G^* and S^* are their free energies and entropies at their respective melting points. If the melt formed by these solids is ideal (hence no heat of mixing), then the free energies of these components in their mutual melt are given by

$$G_{A_{liq}} = G^*_{A_{liq}} - (T - T_A)S^*_{A_{liq}} + RT \ln X_A$$

and

$$G_{B_{liq}} = G^*_{B_{liq}} - (T - T_B)S^*_{B_{liq}} + RT \ln X_B$$

At equilibrium

$$G_{A_{liq}} = G_{A_{sol}}$$

and

$$G_{B_{liq}} = G_{B_{sol}}$$

Since

$$G^*_{A_{liq}} = G^*_{A_{sol}}$$

and

$$G^*_{B_{liq}} = G^*_{B_{sol}}$$

we have

$$RT \ln X_A = (T - T_A)(S^*_{A_{liq}} - S^*_{A_{sol}})$$

and

$$RT \ln X_B = (T - T_B)(S^*_{B_{liq}} - S^*_{B_{sol}})$$

Since $S_{\text{liq}}^* - S_{\text{sol}}^*$ is the entropy of fusion, ΔS_f, the equations can be rewritten to read

$$X_A = \exp\left[-\frac{\Delta S_{fA}(T_A - T)}{RT} \right]$$

and

$$X_B = \exp\left[-\frac{\Delta S_{fB}(T_B - T)}{RT} \right]$$

Thus, at T_A, X_A will equal unity, and at absolute zero X_A will equal 0. If T_A is assumed to be 1000°K, T_B to be 1500°K, and ΔS_{fA} and ΔS_{fB} to be 4 cal/deg-mole,

$$X_A = \exp\left[-\frac{2(1000 - T)}{T} \right]$$

and

$$X_B = \exp\left[-\frac{2(1500 - T)}{T} \right]$$

These curves are plotted in Fig. 10.2.

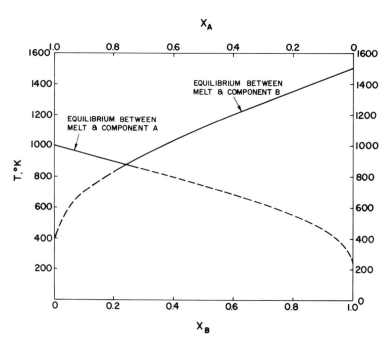

Fig. 10.2. Calculated melting curves for two solids which form a eutectic. The eutectic point occurs at 875°K with a composition of 24.5% component B which has a melting point of 1500°K and 75.5% component A which has a melting point of 1000°K. The dashed lines give "apparent melting points" of the components in mixtures beyond the eutectic point.

Consider what would happen if a melt consisting of 0.95 mole of A and 0.05 mole of B were cooled.   No crystals would appear until a temperature of 975°K was reached.   At this point the curve for equilibrium between the melt and compound A would be reached.   Crystals of component A would form and the melt would become enriched in component B (that is, X_B would rise).   With further cooling the system would move down equilibrium curve A toward its intersection with equilibrium curve B (the eutectic point).   At the eutectic point, B and A would simultaneously crystallize until the liquid disappeared.   Only after the liquid completely disappeared could further cooling take place.   No liquid can stably exist below the eutectic point, 870°K in this case.

Any liquid with more than 75 percent component A would follow a similar cooling history.   Those with less than 75 percent A would first intersect the B equilibrium curve and crystallize component B rather than A until the eutectic point was reached.

Several general principles can be seen from this treatment.   First, because the entropy of fusion is always positive, a lowering of the melting point always takes place when a contaminant which is immiscible in the solid phase is added to the pure melt.   When the contaminant becomes so great as to make the original component negligible in amount, the "apparent" melting point of the component will approach absolute zero. The shape of the equilibrium curve in the ideal case depends on the entropy of fusion.   The initial slope (i.e., for $X_A \cong 1$) is

$$\frac{dT_m}{dX_A} = \frac{RT_A}{\Delta S_{fA}}$$

The smaller the entropy of fusion, the more rapid is the initial temperature drop with the addition of a given amount of component B to the liquid.

In all such situations there exists a minimum temperature at which a melt can form (i.e., that temperature at which the curves intersect). Regardless of the proportions in which the solid components are mixed, the first liquid will have the same composition (that of the eutectic). The so-called eutectic temperature and composition are dependent on T_A, T_B, ΔS_{fA}, and ΔS_{fB}.   In all cases the eutectic temperature must be less than either T_A or T_B.

If, as is the case for most mixed silicate liquids, the solution is not ideal, the effect of the heat of mixing must be included.   Whereas this alters the shape of the melting-point-depression curve, its main features are not affected.

The melting-point-depressing contaminant need not be another solid phase but could instead be a volatile component such as H_2O or CO_2. Although volatiles do not readily dissolve in silicate liquids, if sufficient gas pressure is used, as much as 1 mole of H_2O per mole of silicate can be incorporated. CO_2 appears to be about 10 times less soluble in silicate liquids than H_2O. That this process plays an important role in the earth can be seen as follows: The melting point of most minerals rises about 10 deg/kbar of applied mechanical pressure. If the pressure is exerted via water vapor, the melting point *falls* by about 100°C per kilobar over the first few kilobars of applied pressure. The melting-point lowering induced by the dissolved H_2O molecules greatly overcompensates for the increased pressure.

The minerals orthoclase, albitic plagioclase, and quartz, which dominate granite, have melting points of 1150, 1118, and 1723°C, respectively. The eutectic melting point for a mixture of these three minerals is approximately 1000°C. Under a water pressure of 2 kbars the eutectic point is reduced to approximately 700°C. The effect of contaminants in lowering the temperature at which a silicate liquid can form is clearly of great significance to the study of magmatic processes.

MELTING OF IDEAL SOLID SOLUTIONS

Crystals which consist of a solid solution of two pure minerals, such as olivine which is a mutual solution of Mg_2SiO_4 and Fe_2SiO_4, show a very different pattern of melting. If the melt is again assumed to be ideal, the free energies of the two end-member components in the liquid will be

$$G_{A_{liq}} = G^*_{A_{liq}} - (T - T_A)S^*_{A_{liq}} + RT \ln X_{A_{liq}}$$

and

$$G_{B_{liq}} = G^*_{B_{liq}} - (T - T_B)S^*_{B_{liq}} + RT \ln X_{B_{liq}}$$

If the solid solution is also taken to be ideal, then the free energies of the components in the solid would be given by similar equations:

$$G_{A_{sol}} = G^*_{A_{sol}} - (T - T_A)S^*_{A_{sol}} + RT \ln X_{A_{sol}}$$

and

$$G_{B_{sol}} = G^*_{B_{sol}} - (T - T_B)S^*_{B_{sol}} + RT \ln X_{B_{sol}}$$

At equilibrium for component A,

$$RT \ln X_{A_{sol}} - (T - T_A)S^*_{A_{sol}} = RT \ln X_{A_{liq}} - (T - T_A)S^*_{A_{liq}}$$

Again defining ΔS_f as the entropy of fusion,

$$X_{A_{sol}} = X_{A_{liq}} \exp \left[\frac{\Delta S_{fA}(T_A - T)}{RT} \right]$$

Similarly,

$$X_{B_{sol}} = X_{B_{liq}} \exp \left[\frac{\Delta S_{fB}(T_B - T)}{RT} \right]$$

Since

$$X_{A_{sol}} + X_{B_{sol}} = 1$$

and

$$X_{A_{liq}} + X_{B_{liq}} = 1$$

it is readily shown that

$$X_{A_{liq}} = \frac{1 - \exp[\Delta S_{fB}(T_B - T)/RT]}{\exp[\Delta S_{fA}(T_A - T)/RT] - \exp[\Delta S_{fB}(T_B - T)/RT]}$$

and that

$$X_{B_{liq}} = \frac{1 - \exp[\Delta S_{fA}(T_A - T)/RT]}{\exp[\Delta S_{fB}(T_B - T)/RT] - \exp[\Delta S_{fA}(T_A - T)/RT]}$$

With T_A taken to be 1000°K, T_B to be 1500°K, and ΔS_{fA} and ΔS_{fB} to be

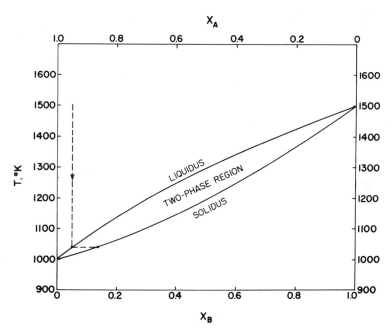

Fig. 10.3. Calculated melting temperature versus mole fraction for two components which form a solid solution. The upper curve gives liquid compositions in equilibrium with solid compositions on the lower curve at any given temperature. The dashed arrow shows the temperature of first crystallizing, and composition of the solid phase which would first crystallize from a liquid containing 95 percent component A.

4 cal/deg, the resulting equilibrium curves are shown in Fig. 10.3. The upper curve represents the loci of liquid-composition–temperature points at which equilibrium can exist between a liquid and solid phase, and the lower curve represents the loci of solid-composition–temperature points at which this equilibrium can exist. Above the upper curve the material is molten, and below the lower curve the material is solid. Points lying between the two curves represent two phase systems of liquid and solid in equilibrium.

If a 95% A–5% B mixture is cooled from 1500°K, the first solid will appear at 1045°K. Its composition will be 86% A and 14% B. Upon further cooling, if equilibrium is maintained, the composition of both the liquid and the solid will move toward A until finally, as the last drop of liquid crystallizes, the solid will have the same composition as the initial liquid.

This calculation allows some generalizations to be made regarding systems involving ideal solid solutions. The solid forming at any given time will always have a greater percentage of the high-melting-point component than the crystallizing liquid. The separation between the liquidus (upper curve) and solidus (lower curve) depends on the magnitude of the fusion entropies. The larger the entropies of fusion, the more separation will there be. Crystallization always takes place at temperatures *between* the melting points of the pure components. Again, nonidealities as reflected by a finite heat of mixing for either the liquid or the solid solution alter the shape of the curves.

INCONGRUENT MELTING

Most solids melt to give a liquid which has the same composition as the original solid. A number of geologically important minerals have the interesting property that on melting they produce a liquid and another solid phase, neither of which has the composition of the original solid. This is known as *incongruent melting*.

Figure 10.4 shows the phase diagram for the system $MgO-SiO_2$ at 1 atm. In this system, clinoenstatite, $MgSiO_3$, is an intermediate compound which melts incongruently. If pure $MgSiO_3$ is heated, a liquid will first appear at 1557°. This liquid will consist (in weight percent) of approximately 39% MgO, 61% SiO_2 whereas $MgSiO_3$ is approximately 41% MgO, 59% SiO_2. Thus the liquid is slightly richer in SiO_2 than the original solid. To compensate for the excess SiO_2 in the liquid, some Mg_2SiO_4 (olivine) forms which has a composition of 42% SiO_2, 58% MgO. After all the clinoenstatite has disappeared, the olivine will begin

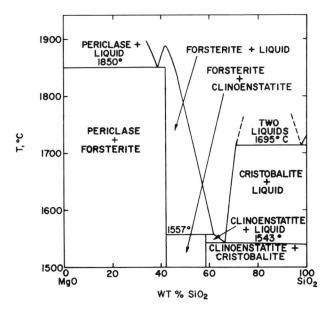

Fig. 10.4. Phase diagram for the system MgO-SiO₂ as determined by Bowen and Andersen (1914) and modified by Grieg (1927).

to melt if the temperature is raised above 1557°C. All the olivine will be melted once the temperature is above approximately 1580°C.

The crystallization of melts whose composition falls between $MgSiO_3$ and Mg_2SiO_4 and between $MgSiO_3$ and the eutectic presents interesting and important cases. If a melt with composition 50% MgO, 50% SiO_2 is cooled, the first solid to form is olivine, Mg_2SiO_4. On further cooling, the liquid becomes more enriched in SiO_2, and more olivine precipitates out of the melt. When the temperature reaches 1557°C, the liquid reacts with the solid olivine crystals to produce clinoenstatite. For the composition chosen, all the remaining liquid would be consumed in this reaction. The system would be completely solidified at 1557°C and would consist of a mixture of olivine and clinoenstatite. If our starting composition had between 61 and 64 percent SiO_2, the first phase to crystallize would have been clinoenstatite. On further cooling, this liquid would continue to directly crystallize out clinoenstatite, with the liquid composition following the equilibrium line down to the cristobalite-clinoenstatite eutectic. This system would become completely solid at 1543°C and would consist of clinoenstatite and cristobalite.

At high pressures the incongruent melting of enstatite disappears, and a eutectic between enstatite and forsterite olivine forms. This is shown schematically in Fig. 10.5. The incongruent melting of enstatite occurs at low pressures, because the melting curve for olivine lies too far above that for enstatite, and no intersection is possible at a composition between Mg_2SiO_4 and $MgSiO_3$. The increase in the melting point of

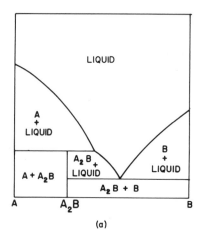

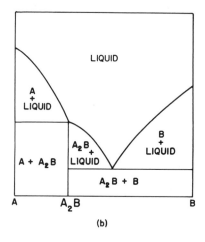

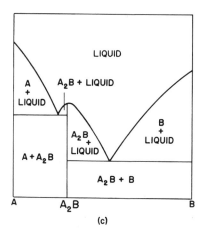

Fig. 10.5. Disappearance of incongruent melting with increasing pressure. *(a)* Incongruent melting of compound A_2B. At some high pressure *(c)* compound A_2B forms eutectics with both A and B. At an intermediate pressure *(b)* the peritectic point passes into the congruent melting point of A_2B.

enstatite with pressure eventually causes the olivine and enstatite melting curves to intersect at an intermediate composition and thus form a normal eutectic.

Earlier in this chapter we showed that the initial slope of the melting curve for a pure end member in a eutectic mixture is given by

$$\frac{dT_m}{dX_A} = \frac{RT_A}{\Delta S_{fA}}$$

This is not true for intermediate compounds which form eutectics with true end members or with other intermediate compounds. It can be shown that the melting-temperature–composition diagram must have a horizontal tangent at the melting point of any intermediate compound. Thus it is impossible to define an initial slope for compounds such as Mg_2SiO_4 or $MgSiO_3$ (in the high-pressure system). For these compounds the entropy of fusion can be obtained from data on systems in which they are not intermediate compounds. For example, it is possible to calculate

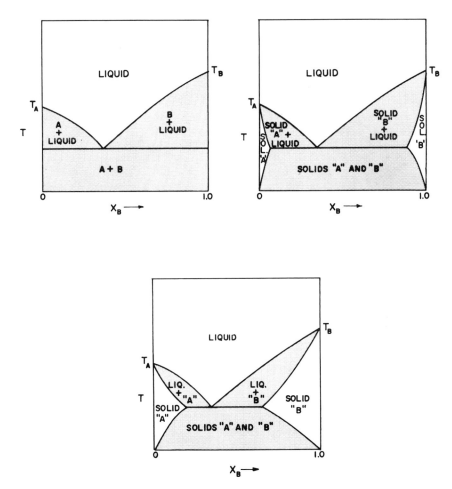

Fig. 10.6. Progression toward increasing miscibility of solids A and B. Solid "A" refers to a solid solution in which A is the major component. White areas indicate conditions under which a single homogeneous phase exists. Shaded areas mark regions where two phases coexist in equilibrium.

the entropy of fusion from the initial slopes of the solidus and liquidus in ideal solid solutions. For Mg_2SiO_4, which forms such a solution with Fe_2SiO_4, the entropy of fusion may be obtained in this manner.

MELTING IN MIXED SILICATE SYSTEMS

Although some mineral pairs closely approximate ideal solid solutions, and most others are almost entirely immiscible, some fall between these extremes. Several mineral pairs, for example, show partial miscibility.

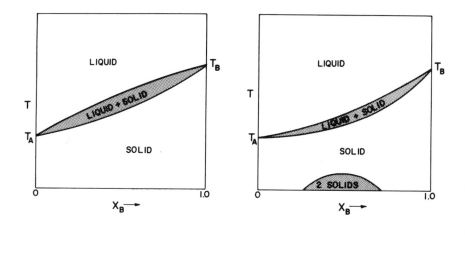

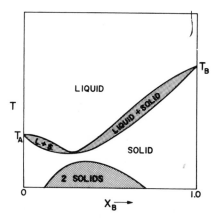

Fig. 10.7. Progression toward decreasing miscibility starting from an ideal solid solution. White area indicates a single homogeneous phase. Shaded areas are two-phase regions.

In other words, small amounts of A can be dissolved in B, and vice versa. A sequence of generalized melting diagrams for increasing miscibility is shown in Fig. 10.6. It can be seen that as long as either solid is in contact with the mixed liquid, the amount of the contaminant it dissolves will increase with decreasing temperature, because the liquid becomes ever richer in the contaminant as the system cools. Once the eutectic temperature is reached and the liquid disappears, the opposite trend ensues. As the two solids cool, their ability to dissolve the contaminant decreases until at absolute zero a mixture of pure A and pure B becomes stable. The reason for this unmixing will be given in the next chapter.

On the other hand, most solid solutions are not ideal. Heat is absorbed during mixing (i.e., the bonds holding the mixture together are not as strong as the average bond strength for the pure end members). This nonideality leads to a downward bulge in the liquidus and solidus curves. If this bulge is sufficiently great to bring the liquidus below both the melting points, the solidus and liquidus must meet at the minimum point. This change due to increasing nonideality is depicted in Fig. 10.7.

As shown in this figure, a nonideality in the solid solution may lead to the additional complication that the solids unmix when cooled to a sufficiently low temperature. At absolute zero, unmixing would be complete and the pure phases A and B would be stable.

It is easy to see how the sequence from complete immiscibility to ideal solid solution is completed. If in Fig. 10.7 the nonideality were increased to the point where the exsolution dome penetrated the liquidus, then the eutectic situation depicted in Fig. 10.6c would result. The greater the degree of overlap, the closer the diagram would approach extreme immiscibility (Fig. 10.6a).

PROBLEMS

10.1 A liquid has a composition of 55 percent by weight MgO and 45 percent SiO_2. Using the MgO-SiO_2 phase diagram, determine the composition and amounts of each phase present after crystallization of 10, 20, 30, 40, 50, 60, and 70 percent of the liquid by equilibrium crystallization. How would your results change if only surface equilibrium were maintained?

10.2 Determine the phase diagram for the following hypothetical material.
Phases:

> Solids, sol A and sol B
> Liquid, liq

Transition parameters:

$$\Delta S^{std}_{sol\ A \to sol\ B} = -2.0 \text{ cal/deg-mole}$$

$$\Delta S^{std}_{liq \to sol\ A} = -36.0 \text{ cal/deg-mole}$$

$$\Delta H^{std}_{sol\ A \to sol\ B} = +200 \text{ cal/mole}$$

$$\Delta H^{std}_{liq \to sol\ A} = -40,000 \text{ cal/mole}$$

$$\Delta V^{std}_{sol\ A \to sol\ B} = -6.0 \times 10^{-3} \text{ liter/mole}$$

$$\Delta V^{std}_{liq \to sol\ A} = -4.0 \times 10^{-3} \text{ liter/mole}$$

Assume ΔS and ΔV remain nearly constant with T and P. Give the coordinates of the triple point and slopes of the three phase boundaries. At what pressure does sol A become stable at room temperatures? What is the 1-atm melting point of sol A? Which phase would be stable at the Moho under a typical continent?

*10.3 Use the results of Prob. 6.2 and the sources of data given there. Additional abundance data required for this problem (from Suess and Urey) are

$$\begin{array}{ll} Si = 10^4 & Mg = 9100 \\ Al = 950 & Ca = 490 \\ K = 32 & Fe = 6000 \\ Na = 440 & \end{array}$$

Are any metallic elements present as solids at 1500°C? At 1200°C? Calculate condensation temperatures for $MgAl_2O_4$, Al_2SiO_5, $CaAl_2Si_2O_8$, Ca_2SiO_4, $CaSiO_3$, $CaMgSi_2O_6$, $KAlSi_3O_8$, $MgSiO_3$, SiO_2, and $MgSiO_4$. What solid phases are present at 1200°C? Remember that previously condensed solids may react with the gas to produce new solids. Will any solid present at 1200°C react with the gas to produce a liquid?

10.4 Dry peridotite melts at 1500°C at 30 kbars. Peridotite mixed with 30 weight percent H_2O begins to melt at 1050°C [I. Kushiro et al., J. Geophys. Res., **73:** 6023 (1968)]. Calculate the mole fraction of H_2O in the hydrous system. Assuming that the depression of the melting point of peridotite depends linearly on the mole fraction of H_2O in the system, plot melting-point depression versus mole fraction of H_2O. A reasonable concentration of H_2O in the mantle is 0.1 percent by weight. Estimate the temperature at which a 2-percent partial melt occurring as a eutectic melt would occur at $P = 30$ kbars, assuming that all the H_2O goes into the melt phase.

*10.5 A volume element of mantle material consisting of 60% olivine, 20% orthopyroxene, 15% clinopyroxene, and 5% garnet is raised adiabatically from 60 to 20 km.   If it is on the solidus at 60 km, how much melting will take place during its ascent?   How will the density change if there are no phase changes other than melting?   *Useful hints:* See Green and Ringwood for the solidus pressure gradient.   Neglect the effect of water.   You may have to estimate some of the parameters, such as compressibility of basaltic liquid.

REFERENCES

Bowen, N. L., and O. Andersen: *Am. J. Sci.*, **37**: 448 (1914).
Boyd, F. R., and J. L. England: *J. Geophys. Res.*, **68**: 311 (1963).
———, ———, and B. T. C. Davis: *J. Geophys. Res.*, **69**: 2101 (1964).
Clark, S. P., Jr., and A. E. Ringwood: *Rev. Geophys.*, **2**: 35 (1964).
Davis, B. T. C., and J. L. England: *J. Geophys. Res.*, **69**: 1113 (1964).
Grieg, J. W.: *Am. J. Sci.*, **13**: 133 (1927).

SUPPLEMENTARY READING

Green, D. H., and A. E. Ringwood: The Genesis of Basaltic Magmas, *Contrib. Mineral. Petrol.*, **15**: 103 (1967).
Ito, K., and G. C. Kennedy: Melting and Phase Relations in a Natural Peridotite to 40 Kilobars, *Am. J. Sci.*, **265**: 519 (1967).
Kushiro, I.: Compositions of Magmas Formed by Partial Zone Melting of the Earth's Upper Mantle, *J. Geophys. Res.*, **73**: 619 (1968).
———, Y. Syono, and S. Akimoto: Melting of a Peridotite Nodule at High Pressures and High Water Pressures, *J. Geophys. Res.*, **73**: 6023 (1968).
Larimer, J. W.: Chemical Fractionations in Meteorites.   I, Condensation of the Elements, *Geochim. Cosmochim. Acta*, **31**: 1215 (1967).
Luedemann, H. D., and G. C. Kennedy: Melting Curves of Lithium, Sodium, Potassium and Rubidium to 80 Kilobars, *J. Geophys. Res.*, **73**: 2795 (1968).
Ringwood, A. E.: Mineralogy of the Mantle, in P. M. Hurley (ed.), "Advances in Earth Science," p. 357, The M.I.T. Press, Cambridge, Mass., 1966.
Schairer, J. F.: Phase Equilibria at One Atmosphere Related to Tholeiitic and Alkali Basalts, in P. H. Abelson (ed.), "Researches in Geochemistry," vol. 2, p. 568, John Wiley & Sons, Inc., New York, 1967.
Yoder, H. S., Jr., and C. E. Tilley: Origin of Basalt Magmas: An Experimental Study of Natural and Synthetic Rock Systems, *J. Petrol.*, **3**: 342 (1962).

chapter eleven

Solid-solution Phenomena

Two minerals with the same basic anion framework but different balancing cations generally show mutual solubility. In some cases, such as Fe and Mg olivines, complete mutual solubility exists and a single homogeneous phase results. In other cases, the solubility is rather limited. For example, in coexisting siderite ($FeCO_3$) and calcite ($CaCO_3$) only a few percent of the Fe will be replaced by Ca in siderite and a few percent of the Ca by Fe in the calcite.

The degree of solid solution depends on the temperature of formation of the solution and on the cooling history of the solid. This is because the degree of mutual solubility decreases with decreasing temperature. A mineral pair which shows complete solubility at the crystallization temperature might exsolve into two separate phases on cooling. Since this exsolution proceeds by diffusion and since diffusion rates fall dramatically with decreasing temperature, there is a temperature below which perceptible exsolution no longer can occur. The extent of total exsolution depends strongly on the rate of cooling from crystallization to the "freeze-in" temperature. Although often the result of a complex interplay between equilibrium and kinetic factors, exsolution phenomena offer the opportunity to investigate the thermal history of a rock.

Most studies of solid-solution phenomena have been empirical.   The phases of interest are mixed and heated to a specified temperature until equilibrium is achieved.   The mixture is then quenched and the degree of solution of phase A in B and of B in A is measured.   This procedure is repeated at a number of temperatures to obtain the solvus curve. An example of a solvus curve in the feldspars is shown in Fig. 11.1. Since the degree of mutual solubility depends on pressure, the solvus must also be determined for a variety of pressures.

In the system $KAlSi_3O_8$-$NaAlSi_3O_8$ at high temperatures, a series of solid solutions is found; any leucite originally formed reacts to form a feldspar as crystallization proceeds.   The alkali feldspar phase diagram (Fig. 11.1) shows the presence of a single homogeneous feldspar over the temperature range from 1050 to about 700°C.   Below 700°C the alkali feldspar does not remain a single phase but "unmixes" to form two feldspars, one of which is rich in Na, the other rich in K.   This unmixing occurs on a scale which can be observed with the aid of a petrographic microscope.   The structure which results from the unmixing of an alkali feldspar is called *perthitic structure.*

The fact that the K and Na feldspars unmix at low temperatures is an example of an important generalization.   If a solid solution is an

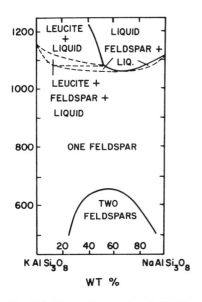

Fig. 11.1. Phase diagram for alkali feldspars as determined by Bowen and
Tuttle (1950) and Schairer (1950) at
1-atm pressure.

ideal solution, the solid phase formed must be homogeneous. Not all homogeneous solutions are ideal, but all inhomogeneous solutions are nonideal. The observation that alkali feldspars unmix at low temperatures shows that the solid solution in the system $NaAlSi_3O_8$-$KAlSi_3O_8$ is nonideal.

The reason for exsolution in the alkali feldspar series can be understood in qualitative terms if we consider the crystal structures of albite and orthoclase. Both these minerals have a framework of Al-Si-O. The two silicate frameworks are similar: the structure for albite is essentially a slightly collapsed version of that for orthoclase. Let us assume we begin with 1 mole of albite-rich feldspar and 1 mole of orthoclase-rich feldspar and then trade K and Na atoms between them. If the two lattices were identical, this could be done without the expenditure of energy. It turns out that the lattice adjustments which take place require a net expenditure of energy. If the trading is done isothermally, this energy will come from the surroundings. The entropy of the surroundings will fall by an amount $\Delta E/T$ (hence by a much larger amount at low temperatures than at high temperatures). The entropy of the crystals will rise as the result of mixing the Na and K atoms; this change will be independent of temperature. Thus we can envision a case where at relatively low temperature the entropy rise in the crystal would be less than the entropy drop in the surroundings. Mixing under these conditions would decrease the entropy of the universe. Hence unmixing rather than mixing would occur. At some higher temperature the entropy drop in the surroundings would just balance the entropy of mixing. Under these conditions the two feldspars would stably coexist. At an even higher temperature the entropy rise of the crystal would exceed the drop in the surroundings, and spontaneous mixing could occur.

The extent of solid solution in the alkali feldspars depends on the energy difference between Na in an albite lattice and Na in an orthoclase lattice, K in an albite lattice and K in an orthoclase lattice, and the free energy decrease resulting from a random distribution of Na and K in a single phase rather than the ordered distribution of ions in two pure phases. In this chapter we will discuss an approach to quantitatively treating this problem, using experimental thermodynamic data. This treatment is made possible by the careful measurement of volumes and heats of solution performed by Waldbaum and the cation-exchange studies of Orville.

FREE ENERGY OF MIXTURES AND SOLUTIONS

When two phases form a mechanical mixture there is no chemical interaction between the phases. The total free energy of a mechanical mixture of A and B is simply given by

$$G_{\text{mech. mix.}} = x_A G_A + x_B G_B$$

where G_A is the free energy of the pure phase A at the temperature and pressure of interest.   For 1 mole of total molecules we may express the same relationship using the chemical potential of the pure phases:

$$G_{\text{mech. mix.}} = x_A \mu_A^{\circ} + x_B \mu_B^{\circ}$$

If we plot the free energy of the mixture as a function of composition, a straight line connecting G_A and G_B is found.   This is the simplest form of a free energy–composition diagram and is shown in Fig. 11.2a.

In an ideal solution the chemical potential of each component is given by

$$\mu_i = \mu_i^{\circ} + RT \ln x_i$$

where μ_i° is the chemical potential of pure i at the temperature and pressure of interest.   Let us consider an ideal solution of two components, A and B, which share a common anion and in which the cations have the same charge.   We may then define 1 mole of solution as the quantity of solution which contains the same number of anions as 1 mole of one of the pure components.   The molar free energy of the solution would be

$$\bar{G}_{\text{soln}} = \sum_i x_i \mu_i$$

or

$$\bar{G}_{\text{soln}} = x_A \mu_A^{\circ} + x_A RT \ln x_A + x_B \mu_B^{\circ} + x_B RT \ln x_B$$

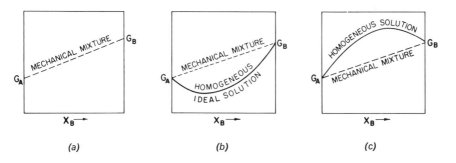

(a)   (b)   (c)

Fig. 11.2. (a) Free energy–composition diagram for a mechanical mixture of components A and B; (b) free energy–composition diagram at 300°K for components A and B which form an ideal homogeneous solution; (c) free energy–composition diagram at constant temperature for components A and B which are totally insoluble.

It is evident that the ideal solution can be related to a mechanical mixture of A and B by

$$\bar{G}_{\text{soln}} = \bar{G}_{\text{mech. mix.}} + x_A RT \ln x_A + x_B RT \ln x_B$$

Since both x_A and x_B are less than 1, both of the logarithm terms will be negative. Consequently an ideal solution of A and B will always have a lower free energy than a mechanical mixture of A and B. This is because a mechanical mixture does not significantly alter the entropy of the system, since the particles being mixed are large. One mole of NaCl consisting of particles 10 microns in size contains only 10^{10} particles but has 10^{23} molecules. Mixing on the molecular scale, such as occurs when an ideal solution is formed, increases the entropy by an amount

$$\Delta \bar{S}_{\text{soln}} = -R(x_A \ln x_A + x_B \ln x_B)$$

Thus the free energy of the solution is lowered. The free energy–composition diagram for an ideal solution is shown in Fig. 11.2b; the mechanical-mixing line is shown for comparison.

From the free energy relationship for an ideal solution we may easily derive the other thermodynamic functions for the ideal solution. The molar volume is found by differentiating the free energy with respect to pressure at constant temperature and composition of the solution:

$$\bar{V}_{\text{soln}} = \left(\frac{\partial \bar{G}_{\text{soln}}}{\partial P}\right)_{T, x_i}$$

$$= x_A \left(\frac{\partial \mu_A^\circ}{\partial P}\right)_{T, x_i} + x_B \left(\frac{\partial \mu_B^\circ}{\partial P}\right)_{T, x_i}$$

$$= x_A V_A^\circ + x_B V_B^\circ$$

where V_A° and V_B° are the molar volumes of pure A and B. Thus there is no volume change when the components mix to form an ideal solution.

The entropy of 1 mole of solution is similarly found by differentiating the free energy with respect to temperature at constant pressure and composition:

$$\bar{S}_{\text{soln}} = -\left(\frac{\partial \bar{G}_{\text{soln}}}{\partial T}\right)_{P, x_i}$$

$$= -x_A \left(\frac{\partial \mu_A^\circ}{\partial T}\right)_{P, x_i} - x_A R \ln x_A - x_B \left(\frac{\partial \mu_B^\circ}{\partial T}\right)_{P, x_i} - x_B R \ln x_B$$

$$= x_A S_A^\circ + x_B S_B^\circ - R(x_A \ln x_A + x_B \ln x_B)$$

The term $-R(x_A \ln x_A + x_B \ln x_B)$ is called the entropy of mixing, because it is the amount by which the entropy of the solution exceeds the entropy of the pure separated components.

The molar enthalpy of the solution is found by using the relationship

$$H = G + TS$$

and the results already found for $\bar{G}_{soln}$ and $\bar{S}_{soln}$:

$$
\begin{aligned}
\bar{H}_{soln} &= (x_A\mu_A^\circ + x_B\mu_B^\circ) + x_A RT \ln x_A + x_B RT \ln x_B \\
&\quad + T[x_A S_A^\circ + x_B S_B^\circ - R(x_A \ln x_A + x_B \ln x_B)] \\
&= x_A(\mu_A^\circ + TS_A^\circ) + x_B(\mu_B^\circ + TS_B^\circ) \\
&= x_A H_A^\circ + x_B H_B^\circ
\end{aligned}
$$

Although both the entropy and free energy of the solution have terms arising from the formation of the solution, the enthalpy of the ideal solution is simply the sum of the contributions of the pure end members.

Some of the properties of an ideal solution are the same as a mechanical mixture of the components. Enthalpy and volume are properties of this type. Thus when two components form a solution for which there is a volume change and an enthalpy change, the solution cannot be ideal.

Some phases which form nonideal solutions at high temperatures unmix completely to a mechanical mixture at low temperatures. This is because at low temperatures the free energy of a homogeneous solution is always greater than the free energy of a mechanical mixture. This case is shown in Fig. 11.2c. If a solution formed at high temperatures is rapidly cooled to a temperature where the solution is unstable, however, the solution may persist if diffusion cannot allow unmixing to occur.

Most solids have properties falling somewhere between those of an ideal solution and those of a system such as shown in Fig. 11.2c. For these solids we may express the free energy of a solid solution of components A and B as

$$
\begin{aligned}
\bar{G}_{soln} = x_A\mu_A^\circ + x_B\mu_B^\circ + x_A RT \ln x_A + x_B RT \ln x_B \\
+ x_A RT \ln \gamma_A + x_B RT \ln \gamma_B
\end{aligned}
$$

where γ_A and γ_B are the activity coefficients of phases A and B in the solution. The real solution described by the equation above is related to the ideal solution by

$$\bar{G}_{soln} = \bar{G}_{ideal} + x_A RT \ln \gamma_A + x_B RT \ln \gamma_B$$

For a real solution the γ's are a function of composition and are usually greater than unity. When γ is greater than 1, the terms with $xRT \ln \gamma$ are positive, and the free energy of the real solution is greater than the free energy of the hypothetical ideal solution of the same composition. The nonideality term opposes the free energy of the ideal mixing term and causes a free energy–composition relationship such as shown by the solid line in Fig. 11.3. The curve has two local minima, which produces the phenomenon of double tangency. A tangent to the curve at composition B_1 also touches the curve at composition B_2. From pure A up to a solution of A and B with $x_B = B_1$, a single homogeneous solution exists. From $x_B = B_2$ to pure B, another homogeneous solution exists. Between $x_B = B_1$ and $x_B = B_2$, two phases coexist; this is because the curve representing a homogeneous solution lies above the double tangent line. The two phases have compositions $x_B = B_1$ and $x_B = B_2$; any composition with $B_1 < x < B_2$ will thus be a mechanical mixture of B_1 and B_2; its free energy will lie on the line of double tangency. The mechanical-mixing lines for pure A and B_1 and for pure B with B_2 are shown by dotted lines. These mixing lines lie above the homogeneous-solution line, indicating that the solution is the stable phase. Since the curve between pure A and B_1 and between pure B and B_2 is concave upward, any mixing line drawn within these segments will lie above the

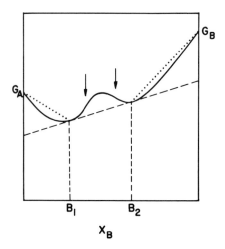

Fig. 11.3. Free energy–composition diagram showing limited solubility of components A and B. Dashed line shows double tangent to the free energy curve. Dotted lines are mechanical-mixing lines for pure A with B_1, and pure B with B_2.

homogeneous-solution curve. Only between B_1 and B_2 will a mixing line fall below the curve.

It is sometimes possible to synthesize solutions with compositions lying between B_1 and B_2. The degree of instability of such solutions is governed by their position with respect to the inflection points on the free energy curve. These inflection points are noted with arrows in Fig. 11.3. For compositions lying between $x_B = B_1$ and the left-hand arrow and between $x_B = B_2$ and the right-hand arrow, the solutions are metastable. Thus they may persist for extended periods of time. Solutions with compositions lying between the two arrows are unstable and will unmix by diffusion, generally in relatively short periods of time.

THE ALKALI FELDSPARS

The alkali feldspars form a geologically important solid-solution series. They are present as major constituents in granitic rocks and in the differentiation products of alkali basalts. The existence of perthite indicates that the solution is extremely nonideal. There are other, more subtle, indications of nonideality in alkali feldspar solutions. For example, there is a positive heat of mixing when the two components combine to form a solution. An ideal solution would have no heat of mixing.

Orville studied the equilibrium between alkali feldspars and 2 m alkali chloride solutions in order to find the extent to which ion exchange between crystals and fluid could occur. His experiments were conducted at 2000-bars pressure and at temperatures between 500 and 700°C. The reaction studied can be written

$$KAlSi_3O_8(xl) + Na^+(fl) = NaAlSi_3O_8(xl) + K^+(fl)$$

The crystals used were sanidine and high-albite structural states. The thermodynamic properties of the feldspars depend on their structural state or the degree of ordering in the Al-Si-O framework, and so it is necessary to have uniform starting materials. At equilibrium,

$$\mu_{Na}{}^{xl} + \mu_K{}^{fl} = \mu_K{}^{xl} + \mu_{Na}{}^{fl}$$

or

$$\mu_{Na}{}^{xl} - \mu_K{}^{xl} = \mu_{Na}{}^{fl} - \mu_K{}^{fl}$$

Thompson and Waldbaum used Orville's data to calculate a theoretical free energy–composition relationship for the alkali feldspar solid solutions.   As discussed above, the free energy for a nonideal solution is

$$\bar{G}_{\text{soln}} = \bar{G}_{\text{ideal}} + x_A RT \ln \gamma_A + x_B RT \ln \gamma_B$$

The activity-coefficient terms may also be written

$$x_A RT \ln \gamma_A + x_B RT \ln \gamma_B = W_A x_A x_B{}^2 + W_B x_B x_A{}^2$$

where W_A and W_B are functions of temperature and pressure but are independent of composition.   Solutions which follow this rule are called regular solutions.   (For a derivation of this relationship see Denbigh.†) The molar free energy of the solid phase is then

$$\bar{G}_{\text{soln}} = \mu_K^\circ x_K + \mu_{\text{Na}}^\circ x_{\text{Na}} + RT(x_K \ln x_K + x_{\text{Na}} \ln x_{\text{Na}}) \\ + W_K x_K x_{\text{Na}}{}^2 + W_{\text{Na}} x_{\text{Na}} x_K{}^2$$

If we perturb the system by removing an infinitesimal amount of $KAlSi_3O_8$ and replacing it by $NaAlSi_3O_8$, the molar free energy will change by

$$\Delta \bar{G}_{\text{soln}} = \mu_{\text{Na}} - \mu_K$$

But the change in G caused by changing the composition by an infinitesimal amount is simply the partial molal free energy of the solution:

$$\Delta \bar{G}_{\text{soln}} = \left(\frac{\partial \bar{G}_{\text{soln}}}{\partial x_{\text{Na}}} \right)_{T,P}$$

Thus

$$\mu_{\text{Na}} - \mu_K = \left(\frac{\partial \bar{G}_{\text{soln}}}{\partial x_{\text{Na}}} \right)_{T,P} = \frac{\partial}{\partial x_{\text{Na}}} \{ (1 - x_{\text{Na}})\mu_K^\circ + x_{\text{Na}}\mu_{\text{Na}}^\circ \\ + RT[(1 - x_{\text{Na}}) \ln (1 - x_{\text{Na}}) + x_{\text{Na}} \ln x_{\text{Na}}] \\ + W_K x_{\text{Na}}{}^2 (1 - x_{\text{Na}}) + W_{\text{Na}} x_{\text{Na}} (1 - x_{\text{Na}})^2 \}$$

$$= \mu_{\text{Na}}^\circ - \mu_K^\circ + RT \ln \frac{x_{\text{Na}}}{x_K} + W_{\text{Na}} \\ + (2W_K - 4W_{\text{Na}})x_{\text{Na}} + 3(W_{\text{Na}} - W_K)x_{\text{Na}}{}^2$$

The free energy of the fluid phase is given by

$$\mu_{\text{Na}} - \mu_K = \mu_{\text{Na}}^* - \mu_K^* + RT \ln m_{\text{Na}}\gamma_{\text{Na}} - RT \ln m_K \gamma_K$$

† K. Denbigh, "The Principles of Chemical Equilibrium," p. 430 ff., Cambridge University Press, New York, 1964.

where μ^* is the chemical potential of the pure component at the temperature and pressure of interest, using the solute standard state. In this case the standard state is defined in terms of molality so that molal concentrations (m) in the fluid phase may be used. (A 1 m solution contains 1 mole of solute in 1000 g of solvent.)

Combining the results for the solid and fluid phases,

$$\mu_{Na}^{\circ} - \mu_{K}^{\circ} + RT \ln \frac{x_{Na}}{x_{K}} + W_{Na} + (2W_{K} - 4W_{Na})x_{Na}$$
$$+ 3(W_{Na} - W_{K})x_{Na}^{2} = \mu_{Na}^{*} - \mu_{K}^{*} + RT \ln \frac{m_{Na}}{m_{K}} + RT \ln \frac{\gamma_{Na}}{\gamma_{K}}$$

Rearranging terms gives

$$RT \ln \frac{x_{K}m_{Na}}{x_{Na}m_{K}} = (\mu_{Na}^{\circ} - \mu_{K}^{\circ} - \mu_{Na}^{*} + \mu_{K}^{*}) + RT \ln \frac{\gamma_{K}}{\gamma_{Na}}$$
$$+ W_{Na} + (2W_{K} - 4W_{Na})x_{Na} + 3(W_{Na} - W_{K})x_{Na}^{2}$$

This expression is a quadratic equation of the form

$$y = a_0 + a_1 x_{Na} + a_2 x_{Na}^{2}$$

where

$$y = RT \ln \frac{x_{K}m_{Na}}{x_{Na}m_{K}}$$

$$a_0 = (\mu_{Na}^{\circ} - \mu_{K}^{\circ} - \mu_{Na}^{*} + \mu_{K}^{*}) + RT \ln \frac{\gamma_{K}}{\gamma_{Na}} + W_{Na}$$

$$a_1 = 2W_{K} - 4W_{Na}$$

$$a_2 = 3(W_{Na} - W_{K})$$

The last two equations can be solved to give values of W_{K} and W_{Na} once a_1 and a_2 are known:

$$W_{K} = -\tfrac{1}{2}a_1 - \tfrac{2}{3}a_2$$
$$W_{Na} = -\tfrac{1}{2}a_1 - \tfrac{1}{3}a_2$$

Thompson and Waldbaum calculated values of y from Orville's ion-exchange data and fitted the ion-exchange data at each temperature to the polynomial equations. Then they calculated values of W_{K} and W_{Na} for each temperature. The result gives the free energy–composition relationship for 2000-bars pressure and temperatures equal to 500, 600, 650, 670, 680, and 700°C. In order to extend the equation of state to other P, T conditions, more data are needed.

The extension of the free energy relationship to other P, T conditions can most conveniently be made if we use the concept of excess functions. The excess function for a thermodynamic variable, Y, is defined as the value of Y for the real solution minus the value for an ideal solution:

$$\bar{Y}_{xs} = \bar{Y}_{real} - \bar{Y}_{ideal}$$

By this definition $\bar{Y}_{xs}$ contains all the nonideal behavior for the variable Y. In the alkali feldspar problem the following excess functions are useful:

$$\bar{G}_{xs} = \bar{G}_{soln} - (\mu^\circ_{Na}x_{Na} + \mu^\circ_K x_K + x_{Na}RT \ln x_{Na} + x_K RT \ln x_K)$$

$$\bar{V}_{xs} = \bar{V}_{soln} - (x_{Na}V^\circ_{Na} + x_K V^\circ_K)$$

$$\bar{S}_{xs} = \bar{S}_{soln} - (x_{Na}S^\circ_{Na} + x_K S^\circ_K - Rx_{Na} \ln x_{Na} - Rx_K \ln x_K)$$

$$\bar{H}_{xs} = \bar{H}_{soln} - (x_{Na}H^\circ_{Na} + x_K H^\circ_K)$$

$$\bar{E}_{xs} = \bar{E}_{soln} - (x_{Na}E^\circ_{Na} + x_K E^\circ_K)$$

We may manipulate the excess functions in exactly the same way as the standard thermodynamic functions. For example,

$$\left(\frac{\partial \bar{G}_{xs}}{\partial T}\right)_{P,x} = \frac{\partial(\bar{G}_{soln} - \bar{G}_{ideal})}{\partial T}$$

$$= -\bar{S}_{soln} + \bar{S}_{ideal}$$

$$= -\bar{S}_{xs}$$

Similarly,

$$G_{xs} = H_{xs} - TS_{xs}$$

$$S_{xs} = \frac{H_{xs} - G_{xs}}{T}$$

Each of the excess functions may be written in a symmetric form, as was done for free energy above:

$$\bar{V}_{xs} = W_{V_{Na}}x_{Na}x_K^2 + W_{V_K}x_K x_{Na}^2$$

$$\bar{H}_{xs} = W_{H_{Na}}x_{Na}x_K^2 + W_{H_K}x_K x_{Na}^2$$

$$\bar{S}_{xs} = W_{S_{Na}}x_{Na}x_K^2 + W_{S_K}x_K x_{Na}^2$$

$$\bar{E}_{xs} = W_{E_{Na}}x_{Na}x_K^2 + W_{E_K}x_K x_{Na}^2$$

The constants W_S, W_V, and W_E are assumed to be independent of temperature and pressure. This is equivalent to saying that there is no nonideal contribution to heat capacity, compressibility, and thermal expansion. The free energy constant, W, is a function of temperature and pressure; the enthalpy constant, $W_H = W_E + PW_V$, is a function of only pressure. Ultimately, the free energy equation will be written in terms of W_E, W_V, and W_S to avoid the problems of variation of the "constants" with temperature and pressure.

The first step in extending the free energy relationship is to find the value of $W_{V_{Na}}$ and W_{V_K}. Since

$$\frac{\partial G}{\partial P} = V$$

this will allow us to extend the equation of state to other pressures. Waldbaum measured the volume of a series of microcline–low-albite feldspars. His results are shown in Fig. 11.4. The upper curve shows

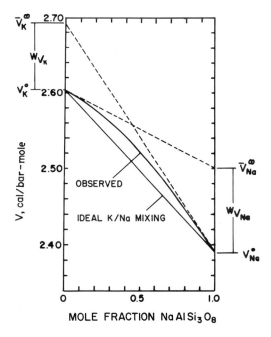

Fig. 11.4. Molar volumes of synthetic microcline–low-albite crystalline solution as a function of composition (upper curve). Straight line below gives theoretical values for an ideal solution. The dashed lines show the graphical evaluation of W_V. To obtain volume in liters per mole, divide by 23.9; the units of calories per bar-mole are suitable for direct use in the free energy equation. (*After Waldbaum.*)

the best fit of a quadratic equation to the volume data. The straight line below gives the expected value for an ideal solid solution. The tangents to the actual volume curve where it reaches pure Na and where it reaches pure K feldspar are also shown. These tangents yield the Henry's law coefficients. Their extrapolation to an intersection on the opposite pure end member yields the partial molar volume $\bar{V}_K^\infty$ of a small amount of microcline added to pure albite and $\bar{V}_{Na}^\infty$ of small amounts of albite added to microcline. (Microcline is the form of $KAlSi_3O_8$ in which Al and Si are perfectly ordered.) The differences between these Henry's law molar volumes and the actual molar volumes of pure albite and microcline are W_{V_K} and $W_{V_{Na}}$.

The use of a symmetric quadratic form for the excess functions can be qualitatively understood by closer examination of the volume curve. If we were to start with pure microcline and begin to replace K with Na atoms, over the first few percent the excess volume would be approximately

$$\bar{V}_{xs} = x_{Na} W_{V_{Na}}$$

However, an excess calculated in this way would soon begin to exceed the observed value. A better fit would be obtained if instead we used the equation

$$\bar{V}_{xs} = x_{Na} W_{V_{Na}} x_K^2$$
$$= x_{Na} W_{V_{Na}} (1 - x_{Na})^2$$

This in effect takes into account the fact that the partial molar excess volume contributed by albite decreases as the lattice becomes more albitic and that this decrease proceeds as the square of the albite content. This remodeled equation would bend our curve down and allow it to follow the experimental curve a bit further. Again, a deviation would set in: this time the curve would be on the low side. The problem now is that we have neglected the contribution of K feldspar to the excess volume as the lattice becomes less like pure microcline. This can be remedied by adding a corresponding term for microcline. Hence,

$$\bar{V}_{xs} = x_{Na} x_K^2 W_{V_{Na}} + x_K x_{Na}^2 W_{V_K}$$

This equation yields a good fit to the experimental data. The values of the coefficients (W_V) calculated for the feldspar system are

$$W_{V_K} = 0.0888 \pm 0.0048 \text{ cal/bar-mole}$$
$$W_{V_{Na}} = 0.1114 \pm 0.0048 \text{ cal/bar-mole}$$

For sanidine–high-albite feldspars, Waldbaum estimated

$$W_{V_{Na}} = 0.0787 \pm 0.0048 \text{ cal/bar-mole}$$

$$W_{V_K} = 0.0787 \pm 0.0048 \text{ cal/bar-mole}$$

The extension of the free energy equation to variable temperature requires calculation of $W_{S_{Na}}$ and W_{S_K}, since

$$\frac{\partial G}{\partial T} = -S$$

From the relationship

$$\bar{S}_{xs} = \frac{\bar{H}_{xs} - \bar{G}_{xs}}{T}$$

we have

$$W_{S_{Na}} = \frac{W_{H_{Na}} - W_{Na}}{T}$$

$$W_{S_K} = \frac{W_{H_K} - W_K}{T}$$

The values of W_{Na} and W_K are known at six temperatures. Knowledge of $W_{H_{Na}}$ and W_{H_K} at these temperatures would permit calculation of $W_{S_{Na}}$ and W_{S_K}.

To evaluate W_H, Waldbaum measured the heats of solution for alkali feldspars of various compositions in HF. His results are shown in Fig. 11.5. The upper curve is the best fit to the experimental data, using a quadratic equation. The lower line is the expected result for an ideal solid solution. W_H may be found by extending the Henry's law slope from infinite dilution up to unit mole fraction and subtracting $\Delta H°$, the heat of solution of the pure phase. For example,

$$W_{H_K} = \Delta \bar{H}_K^{\infty} - \Delta H_K°$$

The values found were

$$W_{H_{Na}} = 8426 \pm 43 \text{ cal/mole}$$

$$W_{H_K} = 6244 \pm 43 \text{ cal/mole}$$

The values of W_E may be calculated immediately, since

$$W_E = W_H - PW_V$$

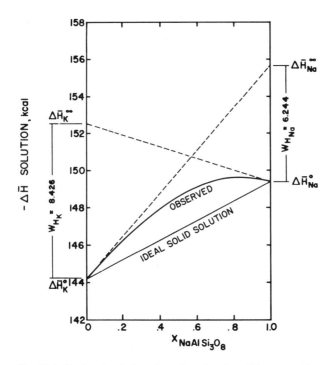

Fig. 11.5. Heats of solution for microcline–low-albite crystalline solutions, showing least-squares fit to data. Lower line shows heats of solution for mechanical mixtures. The dashed lines show graphical evaluation of W_H. *(After Waldbaum.)*

At 1 atm, where the heat-of-solution data were obtained, PW_V is negligible and

$$W_E \cong W_H$$

For sanidine–high-albite solutions there are no direct heat-of-solution data from which W_H may be calculated. However, by comparing W_H for microcline–low-albite solutions with W_H for NaCl-KCl solutions, Waldbaum concluded that W_H (sanidine) $\cong W_H$ (microcline) and W_H (high albite) $\cong W_H$ (low albite). At 2000 bars, $W_{H_K} = 8583$ cal/mole for sanidine and $W_{H_{Na}} = 6401$ cal/mole for high albite. The values of W_S found by using the expression $W_S = (W_H - W)/T$ for each of the end members were

$$W_{S_K} = 4.484 \text{ cal/deg-mole} \quad \text{sanidine}$$

$$W_{S_{Na}} = 3.702 \text{ cal/deg-mole} \quad \text{high albite}$$

The values found so far are tabulated in Table 11.1. Unfortunately, no data are available to calculate W_S for microcline–low-albite solutions.

TABLE 11.1†

Constant for Excess Function at 1 atm	Microcline K	Low Albite Na	Sanidine K	High Albite Na
W_V, cal/bar-gfw	0.0888	0.1114	0.0787	0.0787
W_H, cal/gfw	8426	6244	(8426)	(6244)
W_E, cal/gfw	8426	6244	(8426)	(6244)
W_S, cal/deg-gfw	....	....	4.484	3.702

† Data from D. R. Waldbaum, Unpublished Ph.D. thesis, Harvard University, Cambridge, Mass., 1966. More recent data on Sanidine–high-albite solutions are given in: J. B. Thompson, Jr., and D. R. Waldbaum, Mixing Properties of Sandine Crystalline Solutions III, *Amer. Mineral.*

We now have a complete equation for the free energy for sanidine–high-albite solutions:

$$G = \mu^\circ_{Na}x_{Na} + \mu^\circ_K x_K + RT(x_{Na}\ln x_{Na} + x_K\ln x_K) + (W_{E_{Na}} + PW_{V_{Na}} - TW_{S_{Na}})x_{Na}x_K^2 + (W_{E_K} + PW_{V_K} - TW_{S_K})x_K x_{Na}^2$$

The first three terms are the ideal mixing terms:

$$G_{ideal} = \mu^\circ_{Na}x_{Na} + \mu^\circ_K x_K + RT(x_{Na}\ln x_{Na} + x_K\ln x_K)$$

The last two terms arise from the nonideality in the solution and are equivalent to terms involving the activity coefficients,

$$(W_{E_{Na}} + PW_{V_{Na}} - TW_{S_{Na}})x_{Na}x_K^2 + (W_{E_K} + PW_{V_K} - TW_{S_K})x_K x_{Na}^2$$
$$= RT(x_{Na}\ln\gamma_{Na} + x_K\ln\gamma_K)$$

The values of the constants W_E, W_V, and W_S may be found in Table 11.1. However, we have not yet determined the conditions under which a single homogeneous solution is stable. To do this, we must use the free energy–composition diagram. We shall plot the equation for the free energy of a homogeneous solution and then determine under what conditions this solution is more stable than a mixture of two other solutions, one of which is rich in K and the other rich in Na.

Figure 11.6 shows the free energy function for sanidine–high albite evaluated at several constant temperatures. By using this diagram we may determine the fields of stability of the various phases and find the solvus curve needed to determine the subsolidus temperature-composition relationships. At 800°C a single homogeneous solution is stable over the entire composition range. At 691°C there is still one homogeneous phase at every composition. On the 600°C isotherm it is possible to draw a line of double tangency to the free energy curve; at this temperature there

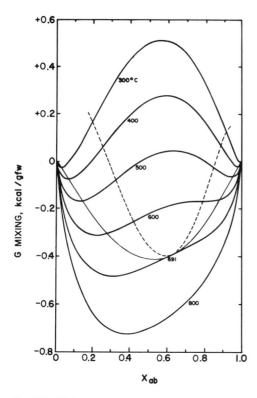

Fig. 11.6. Gibbs free energy of mixing, at constant temperature, as a function of composition, for several temperatures at 2000 bars. Critical temperature at this pressure is 690.3°C and critical composition is 62.3 mole percent $NaAlSi_3O_8$ (ab). The light solid curve gives the binodal curve; the dashed curve is the spinodal curve. See text for further explanation. *(After Waldbaum.)*

is a wide composition range over which the coexistence of two phases is more stable than a single homogeneous solution. The extent of solid solution further decreases as the temperature drops. At 0°K there would be no solution at all, and the $G - x$ curve would be like that shown in Fig. 11.2c.

The solid line joining the points of double tangency to the free energy curve is called the *binodal curve*. The dotted line joining the inflection points on the free energy curve is called the *spinodal curve*. Compositions lying inside the spinodal curve are unstable with respect to diffusion of Na and K. Compositions lying between the spinodal and binodal curves are metastable. Compositions outside the binodal are stable. The binodal and spinodal curves determined in this diagram may be transposed to the more familiar temperature-composition diagram, as shown

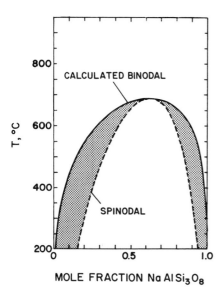

Fig. 11.7. Temperature versus composition projection of binodal and spinodal curves for sanidine–high-albite crystalline solutions at 2000 bars. *(After Waldbaum.)*

in Fig. 11.7. The shaded area lying between the spinodal and binodal curves is the region of a metastable homogeneous solution.

By using Waldbaum's equation of state, it is possible to calculate solvus curves for a series of pressures. His results are shown in Fig. 11.8. The experimental data of Bowen and Tuttle obtained at 981 bars of water pressure are shown for comparison. The agreement between theoretical and experimental data is good. The effect of pressure on solid solutions is clearly seen in this diagram. At high pressures, where the lattice is very compressed, the ability of sanidine and high albite to form a solid solution is reduced. A higher temperature is required to reach a given extent of mutual solubility at high pressure than at low pressure. The temperature at which a single homogeneous phase becomes stable over the entire composition range is called the *critical mixing temperature;* this critical mixing temperature increases by about 9 deg/kbar of pressure.

THE ADVANTAGES OF EQUATIONS OF STATE

Waldbaum's development of an equation of state for the alkali feldspars is an excellent example of the power of thermodynamics to describe complicated physical-chemical systems, using a minimum amount of experimental data. The field of phase equilibrium, particularly concerning subsolidus phases, has long been an experimental field. A system of known bulk composition is subjected to a certain temperature and pressure. The experimenter then examines the resulting mineral phases and their chemical compositions. In this manner a phase diagram can be constructed, usually in the form of a temperature-composition diagram at a fixed pressure. There are many problems associated with this method. First, the data are sometimes very difficult to obtain because of the slow rates of solid-solid reactions. Second, the data cannot be extrapolated to other pressures, since the theoretical relations are not known.

The theory of thermodynamics will not allow us to construct a completely theoretical equation of state for real solids. However, it does allow the formulation of a semiempirical equation of state using a minimum amount of data. In the case of the alkali feldspars, Waldbaum measured the volumes and heats of solutions for alkali feldspars and used existing ion-exchange data to determine the required entropy function. Not only does this method make the maximum use of the minimum amount of data, but also, the results may be extrapolated to other conditions since a complete equation of state was determined.

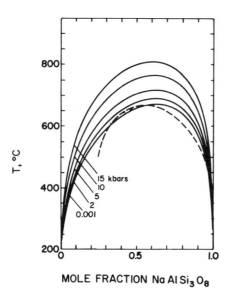

Fig. 11.8. Temperature-composition projection of binodal surface at several pressures for sanidine–high-albite solutions. Dashed curve is solvus (binodal) determined by Bowen and Tuttle (1950) at 981-bars water pressure. *(After Waldbaum.)*

EXSOLUTION OF IRON–NICKEL ALLOYS AND THE COOLING RATES OF IRON METEORITES

The distribution of the elements Ni and Fe in iron meteorites provides a good example of the role of kinetics in exsolution phenomena.   In the early nineteenth century, Widmanstätten noticed that the polished surface of iron meteorites commonly showed a pattern of intersecting bands such as shown in Fig. 11.9.   These bands are actually the surface expression of plates of the metallic phase, kamacite.   The Widmanstätten structure is due to the arrangement of four sets of kamacite plates in an octahedral pattern.   The meteorites which show this Widmanstätten structure are called octahedrites.

Kamacite is a low-nickel alloy, with about 6 percent Ni.   Since iron meteorites have up to 15 percent Ni, the space between kamacite plates must contain a substance with higher Ni content.   Actually, the space is filled by two substances.   A thin rim next to the kamacite is a high-nickel alloy called taenite (Ni = 30 to 50 percent).   The bulk of the

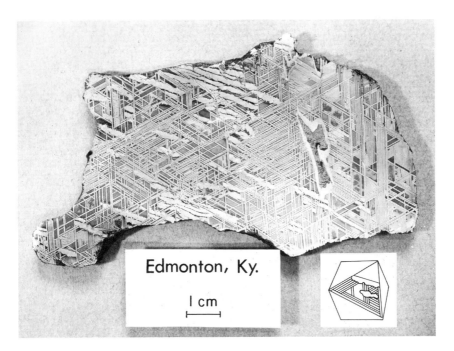

Fig. 11.9. Widmanstätten pattern in the Edmonton, Kentucky, iron meteorite. The bands are formed by the intersection of plates of kamacite with the cut surface of the meteorite. The small reference octahedron shows schematically the orientation of kamacite plates. *(After Wood, 1968.)*

space between kamacite plates is a fine-grained mixture of kamacite and taenite called plessite.

The recent development of the electron microprobe allows investigators to measure the Ni content of the different metallic phases on a very

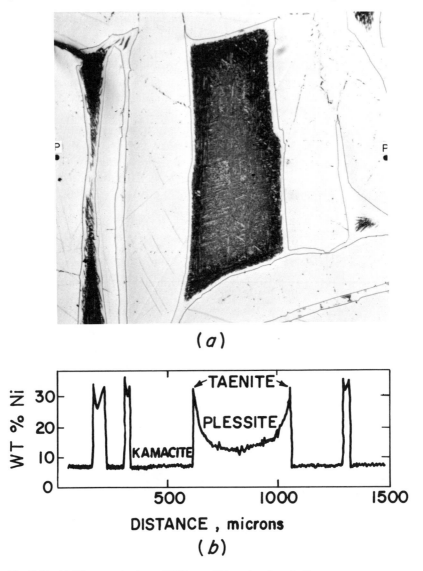

(a)

(b)

Fig. 11.10. (a) Microscopic view of Widmanstätten structure in the Anoka octahedrite, viewed in reflected light. Dark plessite areas are thinly rimmed by taenite; the remaining broad bands of light material are kamacite. (b) Electron-microprobe profile from P to P' showing variation in Ni content. (After Wood, 1964 and 1968.)

small scale.   The microprobe beam can be focused on a spot about 1 micron in diameter.   With a polished thin section of meteorite cut perpendicular to one set of kamacite plates, a series of spot measurements is made along a line which cuts across several kamacite bands and plessite areas.   A typical microscopic view of Widmanstätten structure and the resulting Ni profile are shown in Fig. 11.10.   The M-shaped profiles with irregular interiors are characteristic of the Ni distribution in iron meteorites.   Another feature of the Ni distribution which is always observed is the variation of the magnitude of the Ni minimum in the plessite areas as the width of the M changes.

We may understand the formation of Widmanstätten structure by referring to the phase diagram for Fe-Ni alloys.   Figure 11.11 shows the subsolidus phases for Fe-Ni alloys; these alloys would have crystallized at about 1400°C.   Above 900°C the single homogeneous phase taenite is stable.   Below 900°C kamacite is the stable phase for low Ni contents and taenite for high Ni contents.   A two-phase region exists between the stability fields of kamacite and taenite.   Suppose an iron meteorite with 10 percent Ni was slowly cooled from 900°C.   From 900 to 700°C a single homogeneous phase (taenite) would be present.   At 700°C the meteorite would enter the two-phase (kamacite plus taenite) field, and if equilibrium were maintained, kamacite would begin to grow at the expense of taenite.   In the iron meteorites, kamacite growth was initiated along the octahedral lattice planes of the original taenite crystals.

As the meteorite cools further, the Ni content in both the kamacite and taenite phases must increase.   But since the Ni content in the

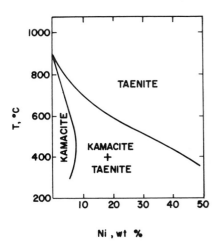

Fig. 11.11. Subsolidus phase diagram for Fe-Ni at 1 atm. *(After Goldstein and Ogilvie.)*

kamacite remains less than the bulk Ni content of the meteorite, the growth of kamacite requires that, during conversion of taenite to kamacite, Ni be pushed out. This excluded Ni enters the adjacent taenite and serves to increase its Ni content as required by the phase diagram. This process continues to 500°C, at which point the equilibrium Ni content for kamacite begins to decrease, whereas that for taenite continues to increase.

The Ni content in kamacite and taenite must continually change if equilibrium is to be maintained. Since the entire system is crystalline, this change must be accomplished by lattice diffusion. Kamacite always must grow at the expense of taenite. This situation is shown schematically in Fig. 11.12.

So far we have assumed that diffusion rates were always fast enough to ensure a homogeneous Ni distribution in both kamacite and taenite phases. But Fig. 11.10b shows that equilibrium was not always maintained. The thin taenite bands have higher Ni contents at their edges than in their centers. In addition, the presence of plessite is due to the failure of diffusion to maintain equilibrium between the kamacite and taenite phases as a whole. We have already shown that diffusion rates decrease very rapidly with decreasing temperature. At high temperatures the diffusion rates maintain a homogeneous Ni distribution in both phases. At lower temperatures diffusion in taenite becomes sufficiently slow to prevent Ni reaching the centers of the taenite crystals. As the temperature further decreases, the central unequilibrated region becomes larger, and the equilibrated edge correspondingly thinner. The unequilibrated centers eventually become sufficiently unstable to break up into a myriad of tiny taenite and kamacite grains (i.e., to plessite).

Kamacite crystals remain equilibrated until below about 500°C.

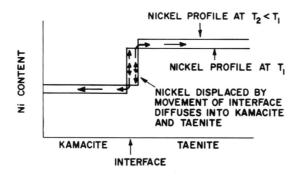

Fig. 11.12. Schematic diagram showing the movement of Ni when kamacite is formed at the expense of taenite, for temperatures above 500°C. *(After Wood, 1968.)*

Below this temperature the kamacite boundary changes slope, and the Ni content starts to decrease. The centers of kamacite crystals show higher Ni content than the edges, indicating that diffusion failed to keep pace with cooling only after the temperature fell beneath 500°C. Wood and several other workers used the diffusion of Ni in taenite to determine the cooling rates of iron meteorites in the temperature range from 600 to 400°C. Using measured diffusion coefficients for Ni in taenite and kamacite, they calculated theoretical Ni profiles for various cooling rates. They found that the Ni distribution in most iron meteorites can be explained by cooling rates of between 1 and 10°C per million years.

The slow cooling rates found for iron meteorites indicate that they must have been shielded by silicate material in their parent bodies. However, the inferred cooling rates are much faster than that for the center of a body the size of the moon. The cooling models calculated by Wood indicate that the iron meteorites were within 100 to 200 km of the surface of their parent "planet."

PROBLEMS

11.1 Calculate the free energy of mixing for sanidine–high-albite solid solutions at 2000-bars pressure and 400, 600, and 800°C. Plot free energy of mixing versus mole fraction of high albite for each isotherm. Give the composition of the stable phases for a system containing 30 mole percent $NaAlSi_3O_8$ at each temperature.

11.2 Calculate the free energy of mixing at 700°C for pressures of 1 bar, 5000 bars, and 10,000 bars for sanidine–high-albite solutions. Plot the free energy of mixing versus mole fraction of high albite for each isobar. Give the composition of the stable phases for a system containing 30 mole percent $NaAlSi_3O_8$ at each pressure.

11.3 Estimate the effect of an error of 10 percent in W_E for sanidine (no error in high albite W_E) on the critical mixing temperature at 1 atm in sanidine–high-albite solutions.

11.4 Solids AX and BX form a nonideal solid-solution series. The heat of mixing ($P = 1$ atm) for AX $= 0.2$ mole fraction is 2.0 kcal/mole and for AX $= 0.8$ is 1 kcal/mole. The volume change on mixing is 0.1 cal/bar-mole at AX $= 0.8$ and 0.08 at AX $= 0.2$. It is known that $W_S(\text{AX}) = 3.5$ cal/deg-mole and $W_S(\text{BX}) = 2.8$. What is the critical mixing temperature at 1 atm?

REFERENCES

Bowen, N. L., and O. F. Tuttle: The System $NaAlSi_3O_8$-$KAlSi_3O_8$-H_2O, *J. Geol.*, **58**: 489 (1950).

Goldstein, J. I., and R. E. Ogilvie: A Re-evaluation of the Iron-rich Portion of the Fe-Ni System, *Trans. Met. Soc. AIME*, **233**: 2083 (1965).

Orville, P. M.: Alkali Ion Exchange Between Vapor and Feldspar Phases, *Am. J. Sci.*, **261**: 201 (1963).

Schairer, J. F.: The Alkali Feldspar Join in the System $NaAlSiO_4$-$KAlSiO_4$-SiO_2, *J. Geol.*, **58**: 512 (1950).

Thompson, J. B., Jr., and D. R. Waldbaum: Mixing Properties of Sanidine Crystalline Solutions. I, Calculations Based on Ion-exchange Data, *Am. Mineralogist*, **53**: 1965 (1968).

Waldbaum, D. R.: Unpublished Ph.D. thesis, Harvard University, Cambridge, Mass., 1966.

Wood, J. A.: The Cooling Rates and Parent Planets of Several Iron Meteorites, *Icarus*, **3**: 429 (1964).

————: "Meteorites and the Origin of Planets," McGraw-Hill Book Company, New York, 1968.

SUPPLEMENTARY READING

Barton, P. B., and P. Toulmin: Phase Relations Involving Sphalerite in the Fe-Zn-S System, *Econ. Geol.*, **61**: 815 (1966).

Goldstein, J. I.: Distribution of Germanium in the Metallic Phases of Some Iron Meteorites, *J. Geophys. Res.*, **72**: 4689 (1967).

*Grover, J. E., and P. M. Orville: The Partitioning of Cations Between Coexisting Single- and Multi-site Phases with Application to the Assemblages: Orthopyroxene-Clinopyroxene and Orthopyroxene-Olivine, *Geochim. Cosmochim. Acta*, **33**: 205 (1969).

Thompson, J. B., Jr.: Thermodynamic Properties of Simple Solutions, in P. H. Abelson (ed.), "Researches in Geochemistry," vol. 2, p. 340, John Wiley & Sons, Inc., New York, 1967.

————: Chemical Reactions in Crystals, *Am. Mineralogist*, **54**: 341 (1969).

———— and D. R. Waldbaum: Analysis of the Two-phase Region Halite-Sylvite in the System NaCl-KCl, *Geochim. Cosmochim. Acta*, **33**: 671 (1969).

———— and ————: Mixing Properties of Sanidine Crystalline Solutions III, *Am. Mineralogist*, **54**: 811 (1969).

chapter twelve **Reactions in Natural Waters**

Many of the chemical transformations which take place within the earth's crust and at its surface involve dissolution by and redeposition from meteoric waters. Rainwater falling on the continents combines with soil acids and dissolves silicate minerals. Much of this dissolved matter is carried to the ocean and is eventually reprecipitated as new mineral matter. In order to understand these processes, it is necessary to define the equilibria between mineral matter and aqueous media. As for all chemical reactions, those involving solutions proceed spontaneously if they lead to an entropy increase for the universe. In order to deal with these reactions, we must obtain free energies for ions in solution.

Dissolved matter in water acts much like the molecules in a gas. Discrete units consisting of from one to ten atoms move about chaotically. When these units are well separated from one another, they act independently. However, when they are packed more closely, mutual interactions complicate the situation. Thus, as for gases, one major source of nonideality stems from electrical interactions between the particles, and the other stems from the occupation by the particles of an appreciable fraction of the available volume.

Unlike gases, the particles in solution need not be electrically neutral; in fact, they are largely in ionic form. Rather than electrical interactions caused by charge asymmetry, it is the net positive or negative charge on the ions which causes the trouble. The higher the charge on the ion, the more complications it creates. As a result, in solutions the degree of nonideality depends not only on the density of the particles but also on their individual charge.

We can handle the effects of nonidealities in aqueous solutions in exactly the same manner we handled them for gases. We will define a dimensionless quantity, a (single-ion activity), such that

$$G_{\text{ion}} = G_{\text{ion}}^{\text{std}} + RT \ln a$$

As for gases, the activity is the ratio of the "effectiveness" of an ion in the solution of interest to its "effectiveness" in a standard solution. The standard state is defined as the activity an ion would have were it at a concentration of 1 mole/liter in a solution free of ion interactions (hence the solution equivalent of an ideal gas). As no real solution with such a high concentration of ions is ideal, the standard state is a hypothetical one. The advantage of this choice is that in solutions dilute enough to be ideal the activity of an ion becomes numerically equal to its concentration in moles per liter.

As was necessary for chemical compounds, we must assign free energies to each species of ion in its standard state. Again no absolute scale is available. If one ion is given an arbitrary value, the free energies of the other ions are uniquely determined. This is most easily seen by considering an example. Solid calcite is equilibrated with water. Solution of the mineral will occur, the reaction being

$$CaCO_3 \rightleftharpoons Ca^{++} + CO_3^{--}$$

The reaction will proceed to the right until the sum of the free energies of Ca^{++} ions and CO_3^{--} ions in solution equals that of the solid calcite:

$$G_{Ca^{++}} + G_{CO_3^{--}} = G_{CaCO_3}$$

Knowing the free energy of calcite, if we were to measure the equilibrium activities of CO_3^{--} and Ca^{++} ions we would be in a position to calculate the sum of the standard free energies for Ca^{++} and CO_3^{--} ions:

$$G_{CaCO_3} = G_{Ca^{++}}^{\text{std}} + RT \ln a_{Ca^{++}} + G_{CO_3^{--}}^{\text{std}} + RT \ln a_{CO_3^{--}}$$

or

$$G_{Ca^{++}}^{\text{std}} + G_{CO_3^{--}}^{\text{std}} = G_{CaCO_3} - RT \ln a_{Ca^{++}} a_{CO_3^{--}}$$

By assigning an arbitrary standard free energy to the CO_3^{--} ion we would fix the free energy of the Ca^{++} ion.

We could repeat this experiment for compounds bearing all the ions of interest. By studying $CaSO_4$ we could obtain the free energy of the SO_4^{--} ion; by studying $MgSO_4$, the free energy of the Mg^{++} ion; and so forth. Instead of defining the free energy of the CO_3^{--} ion, the chemist selected the hydrogen ion. Its standard free energy was designated to be zero. As in the selection of the 92 arbitrary constants mentioned in Chap. 5 necessary to fix the free energies of the solid, liquid, and gaseous compounds, it does not matter what value was chosen for the hydrogen ion. In any calculation we make, the arbitrary value chosen will cancel. Hence the assignment of zero is merely a matter of convenience. In this way our free energy tables can be extended to include species dissolved in water.

Once the standard free energies are in hand, it is easy to deal with reactions for ideal (highly dilute) solutions. Free energy can be expressed in terms of the standard free energy and the numerical value of the concentration:

$$G_i = G_i^{std} + RT \ln [i]$$

Since significant nonidealities set in at very low concentrations, virtually all solutions of interest in earth science are nonideal. Thus we must face the problem of how to relate the thermodynamic property of interest, activity, to the property the chemist measures, concentration.

A quantity γ, the activity coefficient, is defined as the ratio of the actual activity of an ion in the solution of interest to that it would have were the solution ideal. Hence $\gamma = a/a_{ideal}$. Since the ideal activity is numerically equal to the concentration of the ion,

$$\gamma_i = \frac{a_i}{[i]}$$

Our problem is then to establish activity coefficients for the ions of interest as a function of the composition of the solution. The activity of any given ion depends not only on its own concentration in the solution but also on the concentration of all the other ions present. This is because the electrical forces which influence the movements of an ion are generated by all the charged particles in the solution. The degree of nonideality generated by an ion is proportional to the square of its charge. A doubly charged ion creates 4 times the nonideality induced by a singly charged ion. A measure of the nonideality of a solution is thus the sum of the concentrations of each of the ionic species present times the square of its charge.   One-half this summation is defined as the ionic strength, μ, of the solution; hence,

$$\mu = \tfrac{1}{2} \sum_i n_i{}^2 C_i$$

where n_i is the ionic charge and C_i the concentration of the ith species. The factor of $\tfrac{1}{2}$ is introduced so that the ionic strength of a solution of only a monovalent salt (that is, NaCl) is equal to its concentration, the ionic strength of a solution containing only a divalent salt (that is, $CaSO_4$) is 4 times its concentration, and so forth.

Up to ionic strengths in the range of 0.1, the activity coefficient of a given ion has been shown to vary with ionic strength roughly in accordance with the equation

$$\log \gamma_i = -\frac{A n_i{}^2 \sqrt{\mu}}{1 + DB \sqrt{\mu}}$$

where A and B are constants (at 25°C, 0.508 and 0.328, respectively, for aqueous solutions), n_i is the charge on the ion, and D is a constant related to the size of the ion.   Several values of D are given in Table 12.1.   The equation predicts that γ will decrease with increasing μ, approaching the constant value

$$\gamma_i = 10^{-A n_i{}^2 / DB}$$

TABLE 12.1

Constants for Activity-coefficient Equation†

Ion	$D \times 10^8$
$Rb^+Cs^+NH_4{}^+$	2.5
K^+Cl^-	3.0
$OH^-F^-HS^-$	3.5
$Na^+HCO_3{}^-SO_4{}^{--}$	4.2
$Pb^{++}CO_3{}^{--}$	4.5
$Sr^{++}Ba^{++}$	5.0
$Ca^{++}Mn^{++}Fe^{++}$	6
Mg^{++}	8
H^+	9

† Data from I. M. Klotz, "Chemical Thermodynamics," p. 331, Prentice-Hall, Inc., Englewood Cliffs, N.J., 1950.

At 25°C this minimum limit would be approximately 0.1 for Ca^{++} and 0.4 for Na^+. Actually the activity coefficients never reach the limits predicted by this equation. Instead, the trend reverses and they begin to rise toward unity. The minimum activity coefficient for most ions is achieved at about the ionic strength of seawater ($\mu = 0.7$). Curves for several ions are shown in Fig. 12.1.

The mention of experimentally determined activity coefficients for single ions is looked down upon by the purist. In actuality, it is impossible to determine single-ion coefficients. In any experiment only the product of the activity coefficient of one or more positive ions and that of one or more negative ions can be measured. Only by making a special assumption can these products be separated. One way to do this is to assume the activities of K^+ and Cl^- ions are identical (both ions have the same mass, size, and charge). Whereas separation leads to the definite advantage that only values for individual ions need be tabulated (rather than values for each of the innumerable combinations of ions), it has its pitfalls. For the problems considered here we will be safe in using the single-ion coefficients. However, the reader should be cautioned that there are situations where this assumption can lead to serious difficulties.

REACTIONS IN NATURAL WATERS

With this background in mind, we can now consider some of the reactions which take place between ions in natural waters. One type of reaction

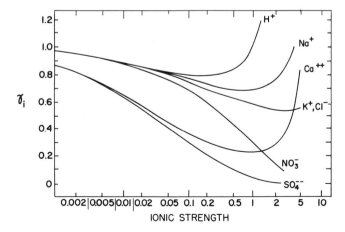

Fig. 12.1. Activity coefficient as a function of ionic strength for single ions. *(After Garrels and Christ, p. 63.)*

dominates all others in importance. Hydrogen (H^+) and hydroxyl (OH^-) ions attach themselves to other ions in the solution. As we shall see, the so-called complex ions which form have a profound effect on chemical reactions taking place between the water and solid material it encounters. Hydrogen ion, of course, tends to combine with negative ions and hydroxyl ions with positive ions.

For example, if the salt Na_2CO_3 is added to water, most of the CO_3^{--} ions which are released will attach themselves to H^+ ions to become bicarbonate ions ($H^+ + CO_3^{--} \rightarrow HCO_3^-$). If a second H^+ ion attaches itself, carbonic acid is formed ($H^+ + HCO_3^- \rightarrow H_2CO_3$). Finally the H_2CO_3 unit can give up both one OH^- and one H^+ ion (hence one H_2O molecule) to become CO_2 gas ($H_2CO_3 \rightarrow CO_2 + H_2O$). If we add a calcite crystal to the solution and then ask ourselves how much calcite will dissolve, we are immediately faced with the problem of determining how much of the carbonate ion from the Na_2CO_3 remains unassociated. The smaller the fraction remaining as carbonate ion, the more solution of calcite will be required to saturate the solution. It is clear that the problem cannot be solved unless the hydrogen complexing is understood.

The first step in deciphering these complexes is to understand the dissociation of H_2O molecules to produce H^+ and OH^- ions. The reaction is

$$H_2O \rightleftharpoons H^+ + OH^-$$

The free energies can be written

$$G_{H^+} = G_{H^+}^{std} + RT \ln a_{H^+}$$

$$G_{OH^-} = G_{OH^-}^{std} + RT \ln a_{OH^-}$$

$$G_{H_2O} = G_{H_2O}^{std} + RT \ln a_{H_2O}$$

At equilibrium,

$$0 = G_{H^+}^{std} + G_{OH^-}^{std} - G_{H_2O}^{std} + RT \ln \frac{a_{H^+} a_{OH^-}}{a_{H_2O}}$$

Since $G_{H^+}^{std}$ is zero and since the activity of H_2O does not change as the result of the very small amount of dissociation (that is, $G_{H_2O} = G_{H_2O}^{std}$ or $a_{H_2O} = 1$) we have

$$RT \ln a_{H^+} a_{OH^-} = G_{H_2O}^{std} - G_{OH^-}^{std}$$

and

$$a_{H^+} a_{OH^-} = \exp \left(\frac{G_{H_2O}^{std} - G_{OH^-}^{std}}{RT} \right)$$

$G_{H_2O}^{std}$ is $-56,700$ cal/mole and $G_{OH^-}^{std}$ is $-37,600$ cal/mole at room temperature, and so the product of the hydrogen- and hydroxyl-ion concentrations must be 1.0×10^{-14}. In order to increase the activity of one ion, that of the other must decrease by an appropriate amount. For a "neutral" solution both concentrations are 1×10^{-7}. If the hydrogen-ion activity is increased to 1×10^{-5}, the hydroxyl activity will fall to 1×10^{-9}.

For convenience the chemist has defined the pH of a solution as follows:

$$pH = -\log a_{H^+}$$

A neutral solution ($a_{H^+} = 1 \times 10^{-7}$) has a pH of 7 and one with a 100-fold higher concentration of H^+ ($a_{H^+} = 1 \times 10^{-5}$), a pH of 5. A pair of electrodes were designed which when immersed in a solution show a difference in electric potential dependent only on the hydrogen-ion activity. Thus the pH (that is, H^+-ion activity) of natural solutions can be directly measured.

If the hydrogen-ion activity in a solution is known, the ratio of the activities of the various forms of dissolved inorganic carbon can be easily calculated. For example, the dissociation of a bicarbonate ion to form a hydrogen ion and a carbonate ion can be written

$$HCO_3^- = H^+ + CO_3^{--}$$

The free energies are

$$G_{HCO_3^-} = G_{HCO_3^-}^{std} + RT \ln a_{HCO_3^-}$$

$$G_{CO_3^{--}} = G_{CO_3}^{std} + RT \ln a_{CO_3^{--}}$$

$$G_{H^+} = RT \ln a_{H^+}$$

At equilibrium,

$$G_{HCO_3^-} = G_{H^+} + G_{CO_3^{--}}$$

Thus

$$-(G_{CO_3^{--}}^{std} - G_{HCO_3^-}^{std}) = RT \ln \frac{a_{H^+} a_{CO_3^{--}}}{a_{HCO_3^-}}$$

and

$$\frac{a_{CO_3^{--}}}{a_{HCO_3^-}} = \frac{1}{a_{H^+}} \exp \left[\frac{-(G_{CO_3^{--}}^{std} - G_{HCO_3^-}^{std})}{RT} \right]$$

Since $G_{CO_3^{--}}^{std}$ is $-126,200$ and $G_{HCO_3^-}^{std}$ is $-140,300$ cal/mole, at room temperature this expression becomes

$$\frac{a_{CO_3^{--}}}{a_{HCO_3^-}} = \frac{5 \times 10^{-11}}{a_{H^+}}$$

In a solution with a pH of 10.3 the activities of HCO_3^- and CO_3^{--} ions would be equal.

The concentration ratio of these two ions is given by

$$\frac{[CO_3^{--}]}{[HCO_3^-]} = \frac{5 \times 10^{-11}}{a_{H^+}} \frac{\gamma_{HCO_3^-}}{\gamma_{CO_3^{--}}}$$

Since doubly charged ions have lower activity coefficients than singly charged ions, the concentration ratio for the two ions is always somewhat greater than their activity ratio.

If every negatively charged ion (anion) in natural waters partially combined with hydrogen, our calculations would become extremely cumbersome. Fortunately, two of the most important species, SO_4^{--} and Cl^- ion, show hardly any tendency to combine with H^+. In natural waters (at room temperature) the concentrations of HSO_4^- and HCl are always negligible compared with those of the uncomplexed forms. Only in concentrated solutions such as seawater does the complexing of another anion with H^+ have to be considered. In such waters the trace element boron is often enriched to the point where it exerts an influence on the pH of the solution. The B present can be in two forms: H_3BO_3 and $H_2BO_3^-$.

In natural waters isolated from the atmosphere, the SO_4^{--} ion often is partially reduced to the S^{--} ion. This reduced S^{--} associates with H^+ ion to form HS^- and H_2S. The partition between these two forms must be considered when dealing with such solutions.

The main cations present in natural waters either have no tendency to complex with OH^- (Na^+, K^+, Mg^{++}, and Ca^{++}) or fully complex with OH^- (Si^{4+}). Thus variable complexing of major cations with OH^- does not present a serious complication in dealing with natural waters. For many trace constituents such as Fe and Mn it is extremely important, however.

FACTORS CONTROLLING THE pH OF SURFACE WATERS

Let us first consider what controls the pH of waters exposed to the earth's atmosphere. Observation has shown that in most cases surface waters have a dissolved CO_2 gas content near equilibrium with the CO_2 partial pressure in the overlying air. Hence the free energy of CO_2 in the water is approximately equal to the free energy of CO_2 in the air (that is, $G^T + RT \ln p_{CO_2}$). As the ratio of CO_2 to $N_2 + O_2$ in the air is remarkably constant, given the elevation and temperature of the water, we can estimate the free energy of its dissolved CO_2. Regardless of whether reactions take place which generate CO_2 or which use up CO_2, exchange with the atmosphere will return the CO_2 gas concentration to this equilibrium value.

A second restriction that can be placed on any surface water is that it must be electrically neutral (i.e., the sum of the negative charges on its anions must exactly balance the sum of the positive charges on its cations).

With these two restrictions we can say a great deal about how the pH of surface waters will vary. We will consider first rainwater, then stream water, and then waters in saline lakes.

Although rainwater contains a small component of entrained sea salt, we can get a good idea of what controls its pH by considering it to be distilled water. CO_2 from the atmosphere dissolves in the raindrops, and the following reaction takes place:

$$CO_2 + H_2O \rightleftharpoons H^+ + HCO_3^-$$

At 25°C the free energies are

$$G_{H_2O} = G_{H_2O}^{std}$$

$$G_{CO_2} = G_{CO_2}^{air} = G_{CO_2}^{std} + RT^{std} \ln p_{CO_2}^{air}$$

$$G_{H^+} = RT^{std} \ln a_{H^+}$$

$$G_{HCO_3^-} = G_{HCO_3^-}^{std} + RT^{std} \ln a_{HCO_3^-}$$

Since at equilibrium $\Delta G = 0$,

$$K_1 = \frac{a_{H^+} a_{HCO_3^-}}{p_{CO_2}}$$

$$= \exp \left(\frac{G_{CO_2}^{std} + G_{H_2O}^{std} - G_{HCO_3^-}^{std}}{RT^{std}} \right)$$

$$= 1.3 \times 10^{-8}$$

As the ionic strength of rainwater is close to zero, the numerical values of the activities of H^+ and HCO_3^- ion will be very nearly the same as their concentrations. Furthermore, as rain will prove to be fairly acid, the concentration of OH^- ion will be so small as not to influence the charge balance (that is, $[HCO_3^-] = [H^+]$). The equilibrium equation can be rewritten

$$K_1 = \frac{[H^+]^2}{p_{CO_2}} = \frac{a_{H^+}^2}{p_{CO_2}}$$

Since K_1 is 1.3×10^{-8} atm^{-1} and p_{CO_2} at sea level is 3.2×10^{-4} atm,

$$a_{H^+} = \sqrt{p_{CO_2}K_1} = 2 \times 10^{-6}$$

Rainwater should then have a pH between 5 and 6.

This water enters the soils, and at some later time it reemerges into a stream. While it is below the ground surface a complex series of reactions occur. CO_2 derived from the bacterial destruction of organic matter enters the water, mineral matter dissolves, ion exchange between the mineral matter and soils takes place, and plants transpire water back to the atmosphere. The net result is that water with a much greater ion content reappears. It contains appreciable amounts of Na^+, K^+, Mg^{++}, Ca^{++}, $Si(OH)_4$, Cl^-, and SO_4^{--}. Once reexposed to the atmosphere, it will again readjust to the dissolved CO_2 content dictated by the atmosphere. For some waters this may involve a release of CO_2 whereas for others an uptake of CO_2 will occur. In any case the pH after equilibration can be calculated in a manner similar to that for rainwater.

The charge-balance equation is

$$[Na^+] + [K^+] + 2[Ca^{++}] + 2[Mg^{++}] = [Cl^-] + 2[SO_4^{--}] + [HCO_3^-]$$

Although H^+, OH^-, and CO_3^{--} ions will be present in the solution, their contributions will be negligible. Again

$$K_1 = \frac{[HCO_3^-](\gamma_{HCO_3^-})(a_{H^+})}{p_{CO_2}}$$

Defining the alkalinity $[A]$ of the solution to be

$$[A] = [Na^+] + [K^+] + 2[Mg^{++}] + 2[Ca^{++}] - 2[SO_4^{--}] - [Cl^-]$$

then in this case

$$[HCO_3^-] = [A]$$

Since in stream waters the ionic strengths are still low enough that the activity coefficients of singly charged ions are close to unity, the hydrogen-ion activity is given by

$$a_{H^+} = \frac{p_{CO_2}K_1}{[A]}$$

Since $p_{CO_2} = 3.2 \times 10^{-4}$ (at sea level) and $K_1 = 1.3 \times 10^{-8}$ (at 25°C),

$$a_{H^+} = \frac{4 \times 10^{-12}}{[A]}$$

The alkalinity which the rainwater inherits during its passage through the soils comes from the solution of carbonate and silicate minerals. For example, when the mineral pyroxene ($MgSiO_3$) dissolves, its Mg becomes a doubly charged ion and its Si becomes a neutral $Si(OH)_4$ molecule. The two plus charges on the Mg are ultimately destined to become balanced by bicarbonate ion. Although the charge balance while the water is in the subsurface could in some cases be established through OH^- ions, as soon as CO_2 becomes available the OH^- will combine with the CO_2 to form HCO_3^-. Similarly, when limestone dissolves, the Ca will go into solution as the double-plus ion and will ultimately be balanced by two HCO_3^- ions. On the other hand, solution of halite ($NaCl$) or gypsum ($CaSO_4$) will not result in an increase in alkalinity. Written in chemical form these reactions are

$$MgSiO_3 + 2CO_2 + 3H_2O \rightarrow Mg^{++} + 2HCO_3^- + H_4SiO_4$$

Pyroxene   Carbon dioxide gas   Water   Dissolved species

$$H_2O + CaCO_3 + CO_2 \rightarrow Ca^{++} + 2HCO_3^-$$

Water   Calcite   Carbon dioxide gas   Dissolved species

$$CaSO_4 \rightarrow Ca^{++} + SO_4^{--}$$

Anhydrite   Dissolved species

$$NaCl \rightarrow Na^+ + Cl^-$$

Halite   Dissolved species

The hydrogen-ion activity of stream waters varies inversely with the quantity of Ca, Mg, Na, and K the waters have derived from weathering carbonate and silicate rocks and with the fraction of the water lost by evaporation. A stream with 4×10^{-5} mole of positive charge to be balanced by HCO_3^- ion will have a pH of 7 and one with 4×10^{-4} mole of positive charge a pH of 8.

The carbonate-ion content of such waters is given by the equilibrium equation

$$K_2 = \frac{a_{H^+}[CO_3^{--}]\gamma_{CO_3^{--}}}{[HCO_3^-]\gamma_{HCO_3^-}}$$

Since $[HCO_3^-] = [A]$, $a_{H^+} = 4 \times 10^{-12}/[A]$, and $K_2 = 5 \times 10^{-11}$,

$$[CO_3^{--}] = \frac{(5 \times 10^{-11})[A]^2\gamma_{HCO_3^-}}{(4 \times 10^{-12})\gamma_{CO_3^{--}}}$$

$$= 12[A]^2 \frac{\gamma_{HCO_3^-}}{\gamma_{CO_3^{--}}}$$

For an alkalinity of 1×10^{-4} mole/liter the carbonate-ion content would be of the order of 10^{-7} mole/liter (hence, as stated above, negligible compared with the major contributors to the charge balance).

The next type of water to be considered is that found in desert lakes. Evaporation over long periods of time has greatly enriched the waters of desert lakes in salt content. In most cases the enrichment has resulted in the precipitation of carbonate and perhaps of silicate minerals. The increase in alkalinity caused by evaporation is thus partially compensated by a loss of alkalinity to these authigenic minerals (i.e., via the inverse of weathering).

Although, as for rain and stream waters, the pH of these lakes is determined by their alkalinities and the p_{CO_2} in the overlying air, several new complications enter into the calculation. First we must consider another type of complex ion. Major cations (Ca^{++}, Mg^{++}, Na^+) associate with major anions (SO_4^{--}, HCO_3^-, CO_3^{--}) to form neutral and monovalent species such as $MgSO_4$, $CaHCO_3^+$, and $NaCO_3^-$. By considering the reaction

$$Na^+ + CO_3^{--} \rightleftharpoons NaCO_3^-$$

we can see why these complexes become more important as the concentration of salt rises. At equilibrium,

$$K_{NaCO_3^-} = \frac{[NaCO_3^-]\gamma_{NaCO_3^-}}{[Na^+]\gamma_{Na^+}[CO_3^{--}]\gamma_{CO_3^{--}}}$$

Since in most solutions there is far more Na^+ than CO_3^{--} ion, the amount of Na ion is not measurably altered by making $NaCO_3^-$ complexes. Hence the ratio of CO_3^{--} in Na complexes to free CO_3^{--} ions is

$$\frac{[NaCO_3^-]}{[CO_3^{--}]} = \frac{K_{NaCO_3^-}\gamma_{Na^+}\gamma_{CO_3^{--}}}{\gamma_{NaCO_3^-}}[\Sigma Na]$$

Since $\gamma_{Na^+} \cong \gamma_{NaCO_3^-}$

$$\frac{[NaCO_3^-]}{[CO_3^{--}]} \cong K_{NaCO_3^-}\gamma_{CO_3^{--}}[\Sigma Na]$$

Clearly the fraction of complexed CO_3^{--} ion will rise with increasing Na content of the solution. Since $K_{NaCO_3^-}$ is about 0.56, the ratio of Na-complexed carbonate to free carbonate ions will reach 0.1 when the Na content of the water reaches 0.9 mole/liter. The existence of other carbonate complexes will strongly affect this calculation, lowering the Na content required to produce a given percent of complexing.

With the exception of K^+ and Cl^-, all the ions present in saline waters participate in major-ion complexes. The job of establishing the fraction of each element in each available complex is a cumbersome one. In seawater, for example, dissolved Mg is partly in the form of Mg^{++} ions, partly as $MgSO_4$, and partly as $MgHCO_3^+$ complexes. Dissolved C occurs as HCO_3^-, $NaHCO_3$, $MgHCO_3^+$, $CaHCO_3^+$, CO_3^{--}, $NaCO_3^-$, $MgCO_3$, $CaCO_3$, and CO_2. Since the various complexes compete with each other for the available ions, a long series of material balance and equilibrium constant equations must be simultaneously solved. In most cases a computer is needed to sort things out. Garrels and Thompson studied a chemical model for seawater. Their results for the degree of complexing of various ions in seawater are shown in Table 12.2.

TABLE 12.2

Seawater Complexes†

	Mole/liter	Free Ion	SO_4^{--} Comp.	HCO_3^- Comp.	CO_3^{--} Comp.
K^+	0.010	99	1	0	0.0
Na^+	0.475	99	1	0	0.0
Mg^{++}	0.054	88	11	1	0.3
Ca^{++}	0.010	91	8	1	0.2
		Free Ion	Na^+ Comp.	Mg^{++} Comp.	Ca^{++} Comp.
Cl^-	0.56	100	0	0	0
SO_4^{--}	0.028	54	21	22	3
HCO_3^-	0.0024	69	8	19	4
CO_3^{--}	0.0003	9	17	67	7

† Data taken from R. M. Garrels and M. E. Thompson, *Am. J. Sci.*, **260**: 57 (1962).

Once the extent of major-ion complexing is established for a given water type, calculations involving variations in trace species (that is, H^+, CO_3^{--}, etc.) can be carried out rather easily. The fraction of free ion can be multiplied by the activity coefficient to yield an apparent activity coefficient. For example, in seawater the fraction of free carbonate ion is 0.10 (the remaining 0.90 resides in Ca, Mg, and Na complexes), and the activity coefficient is 0.20. Hence the apparent activity coefficient is 0.10×0.20, or 0.02. The nonideality of the solution and the incorporation of CO_3^{--} ions into major-ion complexes cause the effectiveness of carbonate ion to be only one-fiftieth that in a hypothetical ideal solution of the same composition and free of major-ion complexes. The apparent and actual activity coefficients for ions in seawater are compared in Table 12.3.

TABLE 12.3

**Apparent and Real Activity Coefficients
for Major Ions in Seawater**

Ion	Real Activity Coefficient†	Apparent Activity Coefficient
K^+	0.64	0.64
Na^+	0.76	0.75
Mg^{++}	0.36	0.32
Ca^{++}	0.28	0.25
Cl^-	0.64	0.64
SO_4^{--}	0.12	0.065
HCO_3^-	0.68	0.47
CO_3^{--}	0.20	0.018

† Activity coefficients from R. M. Garrels and
M. E. Thompson, *Am. J. Sci.*, **260**: 57 (1962).

Another complication which must be faced when dealing with highly saline waters is that CO_3^{--} becomes an important contributor to the charge-balance equation. In this case the alkalinity becomes

$$[A] = [HCO_3^-] + 2[CO_3^{--}]$$

In order to find the hydrogen-ion activity, we must solve this along with two other simultaneous equations:

$$K_1 = \frac{[HCO_3^-]\gamma^*_{HCO_3}a_{H^+}}{p_{CO_2}}$$

and

$$K_2 = \frac{[CO_3^{--}]\gamma^*_{CO_3}a_{H^+}}{[HCO_3^-]\gamma^*_{HCO_3^-}}$$

where γ^* is the product of the activity coefficient and the fraction of the ion free of major-ion complexing.

If we fix the CO_2 partial pressure over a lake and its alkalinity, the three remaining unknowns, HCO_3^-, CO_3^{--}, and a_{H^+}, are fixed. Solving for a_{H^+} we get

$$a_{H^+} = \frac{K_1 p_{CO_2}}{\gamma^*_{HCO_3^-}[A]}\left[\frac{1}{2} + \frac{1}{2}\sqrt{1 + \frac{8(\gamma^{*2}_{HCO_3^-})K_2[A]}{\gamma^*_{CO_3^{--}}K_1 p_{CO_2}}}\right]$$

For alkalinities less than 1×10^{-3} mole/liter, the quantity in the brackets is close to unity. Hence the expression is reduced to that given above for streams. For higher alkalinities the expression in the brackets becomes greater than unity and the hydrogen-ion activity falls more slowly with increasing alkalinity. Using $p_{CO_2} = 3.2 \times 10^{-4}$ atm, $K_2 = 5.0 \times 10^{-11}$, and $K_1 = 1.3 \times 10^{-8}$,

$$a_{H^+} = \frac{4.2 \times 10^{-12}}{\gamma^*_{HCO_3^-}[A]} \left(\frac{1}{2} + \frac{1}{2}\sqrt{1 + \frac{\gamma^{*2}_{HCO_3^-}}{\gamma^*_{CO_3^{--}}} 100 \, [A]} \right)$$

The effect of major-ion complexing is to increase the hydrogen-ion activity (i.e., to make the pH rise more slowly with alkalinity than if major-ion complexes were not present).

The final complication is due to borate. In the ocean and in saline lakes with low alkalinity the HBO_3^- ion contributes significantly to the alkalinity. Hence,

$$[A]^* = [HCO_3^-] + 2[CO_2^{--}] + [H_2BO_3^-]$$

Further, as can be seen from the following reaction, the amount of borate ion depends on pH:

$$H_3BO_3 = H^+ + H_2BO_3^-$$

The equilibrium equation is

$$K_B = \frac{[H_2BO_3^-]\gamma^*_{H_2BO_3^-}a_{H^+}}{[H_3BO_3]}$$

Since

$$[H_3BO_3] = [\Sigma B] - [H_2BO_3^-]$$

we have

$$K_B = \frac{[H_2BO_3^-]\gamma^*_{H_2BO_3^-}a_{H^+}}{[\Sigma B] - [H_2BO_3^-]}$$

or

$$[H_2BO_3^-] = \frac{K_B[\Sigma B]}{K_B + \gamma^*_{H_2BO_3^-}a_{H^+}}$$

The alkalinity $[A]$ to be inserted into the equation yielding a_{H^+} is thus

$$[A] = [A^*] - \frac{K_B[\Sigma B]}{K_B + \gamma^*_{H_2BO_3^-}a_{H^+}}$$

If the total B concentration in the lake and the extent to which B forms major-ion complexes are known, an iterative correction can be made. The presence of B will, in general, raise the hydrogen-ion activity in the lake over that expected if no B were present. The pH will rise more slowly with increasing alkalinity than in the ideal case.

FACTORS CONTROLLING THE OXIDATION STATE OF SURFACE WATERS

In addition to complexing reactions between ions in solutions, electron transfers also take place. Several of the elements dissolved in natural waters can exist in more than one oxidation state. Sulfur can occur in the -2 oxidation state as H_2S or HS^- or in the $+6$ oxidation state as SO_4^{--}. Oxygen can occur in the neutral oxidation state as dissolved O_2 gas or in the -2 oxidation state as H_2O. Hydrogen occurs in the neutral oxidation state as dissolved H_2 gas and in the $+1$ state as H_2O. Carbon occurs in the -4 oxidation state as dissolved CH_4 gas and in the $+4$ state as dissolved CO_2 gas or HCO_3^- ion. Many of the trace metals (Fe, Mn, Cu, U, etc.) found in natural waters also have more than one oxidation state.

Ideally, for surface waters the O_2 content (fixed by the p_{O_2} in the atmosphere) should control the oxidation state of all other elements present. Typical reactions by which this control would be accomplished are

(1) $$O_2 + 2H_2 = 2H_2O$$

(2) $$2O_2 + CH_4 = CO_2 + 2H_2O$$

(3) $$2O_2 + HS^- = SO_4^{--} + H^+$$

(4) $$O_2 + 4Fe^{++} + 4H^+ = 4Fe^{3+} + 2H_2O$$

In terms of the equilibrium constants for these reactions, the following are predicted:

$$p_{H_2} = \sqrt{\frac{1}{p_{O_2}K_1}}$$

$$p_{CH_4} = \frac{p_{CO_2}}{K_2 p_{O_2}{}^2}$$

$$\frac{[HS^-]}{[SO_4^{--}]} = \frac{a_{H^+}\gamma_{SO_4^{--}}}{p_{O_2}{}^2\gamma_{HS^-}K_3}$$

$$\frac{[Fe^{++}]}{[Fe^{3+}]} = \frac{1}{a_{H^+}p_{O_2}{}^{\frac{1}{4}}}\frac{\gamma_{Fe^{3+}}}{\gamma_{Fe^{++}}}\left(\frac{1}{K_4}\right)^{\frac{1}{4}}$$

Since the p_{O_2} and p_{CO_2} are dictated by the atmosphere and the hydrogen-ion activity by the alkalinity and p_{CO_2}, the partial pressures of H_2 and CH_4 in surface waters at sea level and 25°C should be constant. The $[HS^-]/[SO_4^{--}]$ and $[Fe^{3+}]/[Fe^{++}]$ ratios should decrease with increasing pH (hence with increasing alkalinity). It turns out that O_2 gas reacts at a negligible rate with H_2 and with CH_4 at room temperature. These equilibria are never established. Although HS^- and Fe^{++} are oxidized in solutions in contact with the atmosphere, O_2 gas itself does not appear to take part in the reaction. Fortunately, our understanding of the chemistry of surface waters does not depend on the establishment of these equilibria, but as we shall see, the chemistry of subsurface waters is greatly complicated by the sluggishness of oxidation-reduction reactions.

Despite the fact that O_2 gas rarely achieves equilibrium with low-temperature solutions, it is still advantageous to characterize the oxidation state of the species in a solution by the equilibrium oxygen pressure necessary to produce the observed ratios of oxidized and reduced species. For example, were we to measure the ratio of Fe^{3+} to Fe^{++} in a natural water, the equilibrium O_2 gas pressure would be given by

$$p_{O_2} = \frac{1}{K_4} \left(\frac{\gamma_{Fe^{3+}}[Fe^{3+}]}{a_{H^+}\gamma_{Fe^{++}}[Fe^{++}]} \right)^4$$

where

$$K_4 = \exp\left(-\frac{2G_{H_2O}^{std} + 4G_{Fe^{3+}}^{std} - G_{O_2}^{std} - 4G_{Fe^{++}}^{std}}{RT} \right)$$

The oxidation state of a solution can also be defined in terms of an electromotive force (emf) generated between an inert platinum electrode immersed in the solution and a standard calomel reference electrode. The oxidation potential, or so-called Eh, of the solution is related to the generated emf by the following equation:

$$emf = Eh_{soln} - Eh_{calomel}$$

The Eh of a standard calomel electrode is well known, and so the Eh of the solution can be found from a single measurement of emf. This method is customarily used in geologic studies of natural waters. The relationship between the Eh and the equilibrium O_2 partial pressure of the solution is

$$Eh = 1.23 + \frac{0.059}{4} \log p_{O_2} + 0.059 \log a_{H^+}$$

The reader is referred to the text by Garrels and Christ for a complete discussion of problems involving Eh.

DEGREE OF SATURATION OF CALCIUM CARBONATE IN SURFACE WATER

Once the concentrations of the various ionic species in a solution have been defined, it is possible to predict whether the solution is supersaturated or undersaturated with respect to precipitation of a given mineral.

If it is undersaturated, free energy will be released if the mineral dissolves; if supersaturated, free energy will be released if the mineral precipitates. If no free energy change occurs when an infinitesimal amount of the mineral dissolves or precipitates, the solution is exactly saturated with respect to the mineral of interest.

As an example, let us consider the degree of saturation, S, of $CaCO_3$ in surface seawater. S is given by the product of the activities of the Ca^{++} and the CO_3^{--} ions in seawater divided by $k_{S.P.}$, the solubility product for calcite:

$$S = \frac{a_{Ca^{++}} a_{CO_3^{--}}}{k_{S.P.}}$$

where

$$k_{S.P.} = \exp\left(-\frac{G_{CO_3^{--}} + G_{Ca^{++}} - G_{CaCO_3}}{RT}\right)$$

$$= 4.7 \times 10^{-9}$$

Chemical analyses show surface seawater to contain 1.0×10^{-2} mole/liter of Ca. As indicated above, 10 percent of this Ca is tied up in major-ion complexes. The activity coefficient for the free Ca^{++} ions is about 0.28. Hence the Ca-ion activity is 0.25×10^{-2}. Chemical analyses show average surface seawater to contain 2.00×10^{-3} mole of total dissolved inorganic C per liter (that is, $CO_2 + HCO_3^- + CO_3^{--}$). The CO_2 gas content of surface seawater is close to equilibrium with the partial pressure of CO_2 in the atmosphere. From data on the solubility of CO_2 gas in seawater, the corresponding amount of dissolved CO_2 gas is 1×10^{-5} mole/liter. Thus the sum of the HCO_3^- and CO_3^{--} concentrations (all forms including complexes) must be 1.99×10^{-3} mole/liter. The carbonate-ion concentration can be derived from the simultaneous solution of the following three equations:

$$[\Sigma CO_2] - [CO_2] = [HCO_3^-]_T + [CO_3^{--}]_T$$

$$k_1 = \frac{a_{H^+}[HCO_3^-]_T \gamma^*_{HCO_3^-}}{p_{CO_2}}$$

$$k_2 = \frac{a_{H^+}[CO_3^{--}]_T \gamma^*_{CO_3^{--}}}{[HCO_3^-]_T \gamma^*_{HCO_3^-}}$$

Solving in turn for the three unknowns, we get

$$a_{H^+} = 10^{-8.2}$$

$$[HCO_3^-]_T = 176 \times 10^{-5} \text{ mole/liter}$$

$$[CO_3^{--}]_T = 23 \times 10^{-5} \text{ mole/liter}$$

Thus in surface seawater about 12 percent of the dissolved inorganic carbon is in the form of carbonate ion. Of this 12 percent only one-tenth, or 1.2 percent, is uncomplexed. Hence the concentration of free carbonate ions in seawater is 2.0×10^{-5} mole/liter. As the activity coefficient for carbonate ion in seawater is 0.20, the carbonate-ion activity is 0.40×10^{-5}.

Combining the k_{SP} for calcite with the Ca and CO_3 activities for surface seawater, we can calculate the degree of supersaturation:

$$D = \frac{(2.5 \times 10^{-3})(0.40 \times 10^{-5})}{0.47 \times 10^{-8}} = 2.12$$

Thus the surface ocean has a twofold supersaturation with respect to the mineral calcite. The failure of surface seawater to release free energy through the spontaneous precipitation of calcite is just one of innumerable examples of the lack of equilibrium between natural waters and mineral phases at earth-surface temperature.

SUBSURFACE WATERS

The complexity of the chemistry of waters found below the earth's surface becomes apparent if we consider the variations we would encounter along a surface 5 km below sea level. In deep ocean areas this surface would pass through seawater at 1 to 2°C and 500 atm. The CO_2 gas content of this water would be somewhat higher than its surface counterpart and its O_2 content 2 to 4 times lower. On the other hand, beneath continental areas of normal geothermal gradient, the waters encountered would have temperatures between 100 and 200°C and pressures between 500 and 2000 atm. They would, of course, occupy the pore space in the rock. Their compositions would vary widely, more often than not ranging upward to 10 times the salt content of ocean waters. In general, these waters would be O_2-free and in many cases SO_4^{--}-free. Finally, in geothermal areas, waters of even higher temperature would be found. In this chapter we will consider some of the additional information needed in order to cope with chemical reactions taking place in such waters.

TEMPERATURE DEPENDENCES

Clearly, one of the most important matters to consider is the variation with temperature of the equilibrium constants used for aqueous equilibria. We have already seen that the relationship between K and T can be obtained by integrating the expression

$$d \ln K = \frac{\Delta H}{RT^2} dT$$

The ΔH for any given reaction, of course, varies with temperature.

It is convenient to write all reactions so that the products have the higher ionic charge. For example, the dissociation of H_2O may be written

$$H_2O \rightleftharpoons H^+ + OH^-$$

and the solution of a $CaCO_3$ complex

$$CaCO_3 \rightleftharpoons Ca^{++} + CO_3^{--}$$

Any reaction leading to a release of energy during dissociation will have a negative value for ΔH. For such reactions the K and hence the degree of dissociation will decrease with rising temperature. The opposite is true for reactions involving an uptake of energy upon dissociation.

The standard enthalpy changes for reactions of interest show a wide range of values (see Table 12.4). The reason is that the factors controlling the stability of a complex compared with its dissociated counterparts depend in very complicated ways on the electronic structures of the units involved. Thus, as temperatures are raised, the equilibrium constants (and hence the degree of dissociation) move both up and down and at a wide variety of rates. Since typical ΔH values range from 2000 to 20,000 cal/mole, $\ln K$ changes by 0.1 to 1.0 per 10°C temperature change near room temperature.

TABLE 12.4

Thermodynamic Parameters for Several Dissociation Reactions†

Reaction	$\Delta H°$ (1 atm, 25°C)	$\ln K°$
$H_2O = H^+ + OH^-$	13,335	-14.00
$H_2O + CO_2(aq) = H^+ + HCO_3^-$	1,840	-6.35
$H_2S = H^+ + HS^-$	4,800	-6.99
$HCl = H^+ + Cl^-$	$-18,630$	6.1
$HCO_3^- = H^+ + CO_3^{--}$	3,600	-10.32
$HS^- = H^+ + S^{--}$	13,300	-13.90
$HSO_4^- = H^+ + SO_4^{--}$	$-3,850$	-1.99
$MgSO_4 = Mg^{++} + SO_4^{--}$	$-4,920$	-2.25
$CaCO_3 = Ca^{++} + CO_3^{--}$	$-3,130$	-3.20

† Data taken from H. C. Helgeson, *Am. J. Sci.*, **267**: 729 (1969).

Were ΔH to remain constant, the rate at which ln K changed would decrease with rising T.   For example, if a 25°C K doubled for each 10°C rise in temperature, at 325°C (twice the absolute temperature) doubling would occur only once each 40°C.   However, ΔH does not remain constant; the situation is more complicated than this.

To understand how ΔH varies with temperature, we must consider

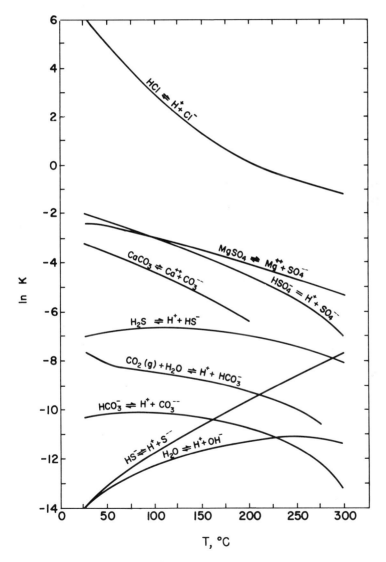

Fig. 12.2. Equilibrium constants for dissociation of complex ions in aqueous solution as a function of temperature.   *(After Helgeson.)*

the heat capacities of ions in solution, since

$$\Delta H = \Delta H° + \int_{T°}^{T} \Delta C_P \, dT$$

As a general rule, the heat capacity of a solution decreases as the result of the dissociation of any complex ions it contains. This is because the electrical interactions of charged species with the water lead, on an average, to higher vibration frequencies. As a result, the ΔC_P for dissociation reactions are negative and their ΔH values fall with increasing temperatures. For those reactions with positive room-temperature ΔH values, a point is eventually achieved where ΔH reaches zero and becomes negative. K (and hence $\ln K$) rises at an ever-decreasing rate until this temperature is achieved and then begins to fall. Since the heat-capacity differences are ordinarily fairly large for dissociation reactions (approximately -50 cal/deg-mole), a reaction with a $\Delta H°$ value of 5000 cal/mole would show a reversal in trend by 125°C. At sufficiently high temperatures, $> 200°C$, almost all complexing reactions favor enhanced association with rising temperature.

In Fig. 12.2 the $\ln K$ values for several important dissociation reactions are plotted as a function of T. Of those reactions shown, only the dissociation of water and of bisulfide ion are more complete at 300°C than at 25°C. In the seven other cases the associated form assumes much greater prominence.

An example will serve to demonstrate just how important these changes are. Let us consider a solution containing 8×10^{-5} mole of $MgSO_4$ per liter. Experience tells us that at room temperature in such a dilute solution there will be a neutral solution (pH = 7) with all the $MgSO_4$ in dissociated form (i.e., as Mg^{++} and SO_4^{--}). Were we to place this solution in a rigid container and heat it to 300°C, complete dissociation would no longer be the case. Mg^{++} and H^+ ions would begin to associate with SO_4^{--} ions to form $MgSO_4^0$ and HSO_4^-. The degree of association could be obtained by solving the following six equations in which six unknowns ($[H^+], [OH^-], [SO_4^{--}], [Mg^{++}], [HSO_4^-],$ and $[MgSO_4^0]$) appear:

$$K_S = \frac{[H^+][SO_4^{--}]}{[HSO_4^-]}$$

$$K_M = \frac{[Mg^{++}][SO_4^{--}]}{[MgSO_4^0]}$$

$$K_W = [H^+][OH^-]$$

$$[\Sigma Mg] = [Mg^{++}] + [MgSO_4^0]$$

$$[\Sigma S] = [MgSO_4^0] + [SO_4^{--}] + [HSO_4^-]$$

$$[H^+] + 2[Mg^{++}] = 2[SO_4^{--}] + [HSO_4^-] + [OH^-]$$

The assumptions have been made that the $MgOH^+$ complex is negligible and that under these dilute conditions activities and concentrations are equal (i.e., all γ's are unity). The results of such calculations carried out at both 25 and 200°C are shown in Table 12.5. At 25°C the solution is indeed neutral and the $MgSO_4^0$ and HSO_4^- complexes negligible. At 200°C half of the Mg is complexed with SO_4^{--} and four-fifths of the SO_4^{--} with Mg^{++} and H^+. The hydrogen-ion concentration has risen by a factor of 1.5 and the hydroxide ion by a factor of 270.

TABLE 12.5

**Solution of $MgSO_4$ at 1 atm, 25°C and 1 atm, 200°C†
(Concentration Units: Moles per Liter)**

Parameter	25°C	200°C
K_W	$10^{-14.00}$	$10^{-11.39}$
K_S	$10^{-1.99}$	$10^{-7.06}$
K_M	$10^{-2.25}$	$10^{-4.8}$
ΣS	8×10^{-5}	8×10^{-5}
ΣMg	8×10^{-5}	8×10^{-5}
$MgSO_4$	1.2×10^{-6}	3.9×10^{-5}
SO_4^{--}	8×10^{-5}	1.5×10^{-5}
HSO_4^-	8×10^{-10}	2.6×10^{-5}
Mg^{++}	8×10^{-5}	4.1×10^{-5}
OH^-	1×10^{-7}	2.7×10^{-5}
H^+	1×10^{-7}	1.5×10^{-7}

† Data for equilibrium constants taken from H. C. Helgeson, *Am. J. Sci.*, **267**: 729 (1969).

The fact that the degree of dissociation of water and bisulfide ion increases steadily with rising temperature whereas that of bicarbonate falls has some interesting consequences. As a solution containing bicarbonate and bisulfide is heated, the OH^- and S^{--} ion concentrations rise at the expense of CO_3^{--}. This enhances the ability of the solution to dissolve carbonates while reducing its capacity to hold oxides and sulfides.

Activity coefficients also change with temperature. These coefficients depend strongly on the dielectric constant of the water. The dielectric constant falls with rising temperature, causing the activity coefficients to decrease. This effect is by no means negligible. For example, in a 2 m NaCl solution at room temperature the product $\gamma_{Na^+}\gamma_{Cl^-}$ is 0.45. At 250°C it drops to 0.15.

Finally, the solubilities of the various mineral phases of interest also change with temperature in accordance with their heats of solution. Although most salts show a positive heat of solution and hence become more soluble with rising temperature, some, like $CaCO_3$, $CaSO_4$, and most silicate minerals, are characterized by negative heats of solution and become less soluble (see Table 12.6).

TABLE 12.6

Thermodynamic Parameters for Several Solution Reactions†

	$\Delta H°$	$\ln K°_{S.P.}$
$CaCO_3 \rightarrow Ca^{++} + CO_3^{--}$	$-3,190$	-8.37
$CaSO_4 \rightarrow Ca^{++} + SO_4^{--}$	$-3,755$	-4.70
$H_2O + CO_2(aq) + ZnS \rightarrow Zn^{++} + HS^- + HCO_3^-$	$10,580$	-18.17
$2CO_2(aq) + 3H_2O + MgSiO_3 \rightarrow$		
$\quad Mg^{++} + 2HCO_3^- + H_4SiO_4$	$-16,375$	-27.99

† Data taken from H. C. Helgeson, *Am. J. Sci.*, **267**: 729 (1969).

PRESSURE DEPENDENCES

Pressure changes of hundreds of atmospheres can result in important changes in the equilibrium constants for dissociation reactions. It can be shown (see Appendix I) that

$$\left(\frac{\partial \ln K}{\partial P}\right)_T = -\frac{\Delta V}{RT}$$

As pressure rises, the equilibrium will gradually shift toward the state which occupies the smallest volume.

In the room-temperature range the ΔV for most dissociation and solution reactions is negative. The dissociated ionic units occupy less space than their associated (or solid) counterparts.

Again let us consider an example. A beaker of water to which 2.00×10^{-4} mole/liter of $Ca(HCO_3)_2$ has been added is allowed to equilibrate with the air. It is then placed in a sealed container and raised to a pressure of 500 atm. We might ask how the degree of saturation, S, of the solution changes when the pressure is applied. Designating the properties of the solution at atmospheric pressure with primes and those at 500 atm with double primes, we can state the problem mathematically as follows:

$$\frac{S''}{S'} = \frac{a''_{Ca^{++}}a''_{CO_3^{--}}/K''_{S.P.}}{a'_{Ca^{++}}a'_{CO_3^{--}}/K'_{S.P.}}$$

We will assume that the activity coefficients of all the ions of interest are unity and that only hydrogen-ion complexes form (i.e., that $CaHCO_3^+$

and $CaCO_3^0$ complexes are negligible). Then

$$a''_{CA^{++}} = a'_{CA^{++}} = \Sigma Ca = 2.00 \times 10^{-4} \text{ mole/liter}$$

and

$$\frac{S''}{S'} = \frac{[CO_3^{--}]''}{[CO_3^{--}]'} \frac{K'_{S.P.}}{K''_{S.P.}}$$

Our problem becomes one of determining how the carbonate-ion content and the calcite solubility change with pressure.

The volume decrease accompanying solution of $CaCO_3$ is about 50×10^{-3} liter/mole. Remembering that at equilibrium at 1 atm

$$0 = G^\circ_{Ca^{++}} + RT \ln a_{Ca^{++}} + G^\circ_{Co_3^{--}} + RT \ln a_{Co_3^{--}} - G^\circ_{CaCo_3}$$

or

$$0 = \Delta G^\circ + RT \ln K'_{S.P.}$$

and since ΔV remains constant over the pressure range of interest, at 500 atm (P'') we have

$$0 = \Delta G^\circ + RT \ln K''_{S.P.}$$
$$= \Delta G^\circ + RT \ln K'_{S.P.} - (P'' - P') \Delta V$$

Combining these two expressions,

$$\frac{K''_{S.P.}}{K'_{S.P.}} = \exp \left[-\frac{(P'' - P') \Delta V}{RT} \right]$$

$$= \exp \left[\frac{(500 \text{ atm}) \times (5 \times 10^{-2} \text{ liter/mole})}{(2 \text{ cal/deg-mole})(300 \text{ deg})(8 \times 10^{-2} \text{ liter-atm/cal})} \right]$$

$$= e^{0.52} = 1.68$$

Thus calcite is 1.68 times more soluble at the higher pressure.

Turning our attention to the carbonate-ion content, we can write the following relationship:

$$K_C = \frac{[CO_2][CO_3^{--}]}{[HCO_3^-]^2}$$

While exposed to the air, the CO_2 content of the solution was at equilibrium with the atmosphere:

$$[CO_2] = \alpha p_{CO_2}$$

In order to balance the charge of the Ca ions,

$$2[CO_3^{--}] + [HCO_3^-] = 2[Ca^{++}]$$

As we shall see, the concentrations of H^+ and OH^- are too low to contribute significantly to the charge balance. Combining these three equations so as to eliminate the CO_2 and CO_3^{--} ion concentrations, we have

$$[HCO_3^-] = \frac{-\alpha p_{CO_2} + \sqrt{\alpha p_{CO_2}(\alpha p_{CO_2} + 16K_C[Ca^{++}])}}{4K_C}$$

By using the values of α, p_{CO_2}, and K_C listed in Table 12.7, the bicarbonate, carbonate, and CO_2 gas content can be calculated. The pH can be evaluated from the expression

$$K_2 = \frac{a_{H^+}[CO_3^{--}]}{[HCO_3^-]}$$

The resulting concentrations are given in Table 12.7.

TABLE 12.7

Dissolution of 2×10^{-4} Moles of $Ca(HCO_3)_2$
(Concentration Units: Moles per Liter)

Parameter	1 atm	500 atm
p_{CO_2}	3.0×10^{-4}	
α	3.4×10^{-2}	
K_C	1.07×10^{-4}	1.15×10^{-4}
K_2	4.8×10^{-11}	5.4×10^{-11}
$[Ca^{++}]$	2.0×10^{-4}	2.0×10^{-4}
$[HCO_3^-]$	4.0×10^{-4}	4.0×10^{-4}
$[CO_3^{--}]$	1.7×10^{-6}	1.5×10^{-6}
$[CO_2]$	1.02×10^{-5}	1.2×10^{-5}
$[\Sigma CO_2]$	4.1×10^{-4}	4.1×10^{-4}
$[H^+]$	1.1×10^{-8}	1.4×10^{-8}
$[OH^-]$	9.0×10^{-7}	8.0×10^{-7}
$[Ca^{++}][CO_3^{--}]$	3.4×10^{-10}	3.0×10^{-10}
$K_{S.P.}$	4.3×10^{-9}	7.2×10^{-9}
S	0.08	0.04

When pressure is applied to the solution and it is no longer free to equilibrate with the atmosphere, its CO_2 pressure can no longer be directly estimated. This restriction is replaced by the requirement that the total amount of dissolved carbon (ΣCO_2) remain the same, provided, of course, that no precipitate forms. Thus we have the three equations

$$[\Sigma CO_2] = [CO_2]'' + [HCO_3^-]'' + [CO_3^{--}]''$$

$$2[Ca^{++}] = [HCO_3^-]'' + 2[CO_3^{--}]''$$

$$K_C'' = \frac{[CO_2]''[CO_3^{--}]''}{[HCO_3^-]''^2}$$

A fourth equation will again give us the pH:

$$K_2'' = \frac{a_{H^+}[CO_3^{--}]''}{[HCO_3^-]''}$$

Before solving these equations, we must estimate by how much the constants K_C and K_2 change with pressure. The volume change for the reaction

$$2HCO_3^- \rightleftharpoons CO_2 + CO_3^{--}$$

is about -8×10^{-3} liter/mole; hence

$$\frac{K_C^{P_2}}{K_C^{P_1}} = \exp\left[\frac{-(P_2 - P_1)\,\Delta V}{RT}\right]$$

$$\frac{K_C^{500}}{K_C^1} = \exp\left[\frac{(500)(8 \times 10^{-3})}{(2)(300)(8 \times 10^{-2})}\right] = e^{0.08} = 1.08$$

For the reaction

$$HCO_3^- \rightleftharpoons CO_3^{--} + H^+$$

the volume decrease is about 11×10^{-3} mole/liter. Hence

$$\frac{K_2^{500}}{K_2^1} = \exp\left[\frac{(500)(11 \times 10^{-3})}{(2)(300)(8 \times 10^{-2})}\right] = e^{0.12} = 1.13$$

By applying these factors to the standard pressure constants and using the [Ca^{++}] and [ΣCO_2] concentrations for the room-air equilibrated sample, the values listed in Table 12.7 are obtained.

COMPLICATIONS RESULTING FROM CHEMICAL INTERACTIONS WITH SOLID PHASES

In nature the waters undergoing changes in temperature and pressure cannot be treated as isolated systems. Transfer of material between these solutions and the solid matter with which they are in contact takes place.   Were such reactions to reach equilibrium, our problem would be greatly simplified.   In most cases the reactions proceed far too slowly for this to be the case.   An excellent example is provided by the interaction between the organic-tissue debris encountered by the water and the dissolved O_2 and SO_4^{--}.   Were thermodynamic equilibrium achieved, either the organic debris or the O_2 and SO_4^{--} would have to be entirely consumed. In most subsurface waters these substances coexist.   The reason is that reaction proceeds at a finite rate only when catalyzed by the enzymatic catalysts found in organisms.   Subsurface waters can be divided into two broad categories on the basis of the manner in which this oxidation proceeds.   Waters in which the O_2 is entirely depleted are referred to as *anaerobic*, and those in which O_2 is only partially used are referred to as *aerobic*.   Their chemistries are vastly different.

AEROBIC WATERS

Deep ocean water provides an example of a situation where a subsurface water undergoes substantial chemical interaction with solid phases without becoming depleted in O_2.

Although it is actually an immensely complicated mechanical and chemical system, the main features of the operation of the ocean can be seen by reducing it to a simple two-layer system.   A relatively thin well-

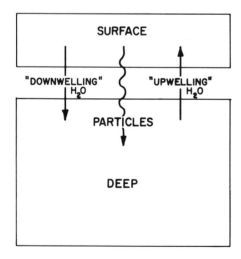

Fig. 12.3. Box model of large-scale ocean mixing.

mixed surface layer at equilibrium with the atmosphere is effectively isolated from a homogeneous deep reservoir which makes up the bulk of the sea (see Fig. 12.3). Water is continuously traded back and forth between these layers. From studies of the distribution of natural radiocarbon it has been shown that, between its relatively brief sojourns at the surface, the average H_2O molecule must reside about 1000 years at depth.

Organisms are formed by photosynthesis from dissolved components in the surface ocean. Their remains (including $CaCO_3$ and SiO_2 hard parts) sink to the deep sea where, to a large extent, they are destroyed by solution and oxidation. The components are returned in dissolved form to the deep ocean. This unidirectional particulate transport causes the chemical composition of deep seawater to differ from that of surface water. This process should lead to the enrichment of the deep sea in Ca, C, and Si with respect to surface water and its depletion in dissolved O_2. These differences are indeed found. Deep water, on an average, contains 1.01 times more Ca, 1.20 times more dissolved inorganic C, and about 100 times more Si than surface water. Deep water shows a threefold depletion in dissolved O_2.

Of the three components falling to the deep sea, only the opal has no effect on the pH of this isolated water. As opal goes into solution largely in the neutral form H_4SiO_4, it does not alter the hydrogen-ion balance. Both the oxidation of organic tissue and the solution of $CaCO_3$ fragments result in pH shifts.

Surface ocean water contains, on an average, about 2.06 moles of dissolved inorganic carbon per cubic meter. This carbon must balance 2.30 moles of positive charge not balanced by the other anions present. To do this requires that 1.80 moles be in the form of HCO_3^- ion and 0.25 in the form of CO_3^{--} ion. The remaining 0.01 mole is in the form of dissolved CO_2 gas. This is the water which enters the deep reservoir. During the 1000 years or so that it remains at depth the typical cubic meter of deep water receives 0.30 mole of CO_2 released by the in situ oxidation of organic debris and 0.10 mole of carbonate ion from the in situ solution of $CaCO_3$. The solution of this $CaCO_3$ raises the amount of positive charge to be balanced by the dissolved inorganic carbon by 0.20 mole/m^3. Thus the water finds itself with a total dissolved inorganic carbon content of 2.46 moles/m^3 (2.06 + 0.30 + 0.10) and a requirement that this carbon balance 2.50 moles/m^3 of positive charge [2.30 + (2 × 0.10)]. In order to maintain electrical neutrality the carbonate-ion content must then be

$$[CO_3^{--}]_{tot} = 2([A] - [\Sigma CO_2])$$
$$= 2(2.50 - 2.46)$$
$$= 0.08 \text{ mole/m}^3$$

This value is 3 times lower than that of 0.25 mole/m^3 for surface water. Except for a small amount of dissolved CO_2, the remaining 2.38 moles will be in the form of bicarbonate ion. Thus the ratio of $[CO_3^{--}]_{tot}$ to $[HCO_3^-]_{tot}$ is slightly more than 3 times smaller in deep than in surface waters.[†] Were the second dissociation constant of carbonic acid the same in the two reservoirs, the hydrogen-ion activity would have to be 3 times higher in deep than in surface water. This can be seen as follows:

$$a_{H^+} = K_2' \frac{[HCO_3^-]_{tot}}{[CO_3^{--}]_{tot}}$$

However, as seen above, the higher pressures in deep water lead to different values for K_2'.

The difference in pH between surface and deep water depends then on (1) the amount of debris returned to solution in a given unit of deep water, (2) the ratio of carbon in the carbonate to that in the organic-tissue form in this debris, (3) differences in the second thermodynamic dissociation constant of carbonic acid resulting from P and T changes, and (4) differences in apparent activity coefficients, γ^*, of CO_3^{--} and HCO_3^- ions due to P and T changes.

ANAEROBIC WATERS

In cases where the oxygen present in the waters is entirely consumed, the oxygen-breathing organisms are replaced by species capable of using SO_4^{--} as an oxidation agent. They continue to consume the available organic fuel. In order to get a feeling for the magnitude of the resulting changes, let us consider a hypothetical ocean identical to the real ocean in all respects except that sulfate rather than oxygen is used to oxidize the organic matter falling into the deep sea. The conversion of 1 mole of SO_4^{--} to S^{--} is equivalent to consuming 2 moles of O_2 gas ($SO_4^{--} \rightarrow S^{--} + 2O_2$). About 1.2 moles of O_2 is required to produce 1 mole of CO_2 via combustion of marine organic tissue. Hence 0.6 mole of sulfate must be reduced to sulfide for each mole of CO_2 generated. Also each mole of sulfide generated increases by 2 moles the amount of positive charge to be balanced by weak bases (HCO_3^-, CO_3^{--}, HS^-, S^{--}). Thus the creation of 0.30 mole of CO_2 per cubic meter of seawater would be accompanied by the generation of 0.18 mole of S^{--} ion and an increase of 0.36

[†] A rigorous formulation of this problem would require a boron correction. Since the rise in hydrogen-ion activity results in a conversion of borate (HBO_3^-) ion to boric acid (H_2BO_3), the resulting decrease in carbonate ion is somewhat smaller than we have calculated.

mole of the positive charge requiring balance by weak anions.   In addition, as before, 0.10 mole of $CaCO_3$ is dissolved in each cubic meter, raising the total CO_2 content by 0.10 and the charge to be balanced by 0.20 mole.   The question is, By how much will these changes alter the carbonate ion and pH of the deep water?

As the second dissociation constant of H_2S is very small, the amount of S remaining in the S^{--} form will be negligible.   It will be partly in the form of H_2S and partly in the form of HS^-.   Thus

$$[H_2S] + [HS^-] = 0.18 \text{ mole/m}^3$$

Assuming that the amount of dissolved CO_2 gas is negligible,

$$[HCO_3^-] + [CO_3^{--}] = 2.46 \text{ moles/m}^3$$

Finally, if the solution is to be neutral,

$$[HS^-] + [HCO_3^-] + 2[CO_3^{--}] = 2.86 \text{ moles/m}^3$$

Simultaneous solution of these equations yields

$$[CO_3^{--}] = 0.22 + [H_2S]$$

Depending on the proportionation of S between the H_2S and HS^- form, the carbonate-ion concentration could range from 0.22 to 0.40 mole/m³. In this case the ratio of carbonate to bicarbonate ion will be nearly equal to or greater than that in surface water and the hydrogen ion will be roughly equal or lower than that in surface water.   Thus the use of sulfate rather than dissolved oxygen gas as an oxidizing agent leads to no pH change or a pH change of the opposite sense.

Most surface waters undergo sufficiently rapid vertical mixing so as to remain aerobic.   However, especially in a situation where a more saline water is overlain by a less saline water (the Black Sea, Dead Sea, certain fiords, lakes fed by saline ground waters, etc.), mixing cannot keep up with the demand for oxidizing agents, and anaerobic conditions result.

For waters trapped in the pores of sedimentary rocks the anaerobic situation is much more common.   The replenishment of O_2 through convection or diffusion cannot keep pace with the activity of organisms.

Because the sulfides of most heavy metals (Fe, Pb, Cu, Zn, etc.) are extremely insoluble, they are rapidly precipitated under anaerobic conditions.   The occurrence of these sulfides in shales can be attributed to the large amount of organic debris incorporated into such sediment, the slow rates at which water flows through such fine-grained material, and the availability of heavy metals. Where anaerobic conditions prevail in metal-poor sediment, such as limestones or evaporites, native sulfur forms rather than metallic sulfides.

The oxidation of organic material is, of course, not the only reaction with the adjacent sediment which affects the chemistry of subsurface waters. Solution and the precipitation of a whole host of mineral phases can be important. As a result, each subsurface water becomes a special case. Its chemistry depends on a complex history of additions and subtractions which have altered its composition since its departure from the surface.

BRINES RECOVERED FROM SEDIMENTS

Most deep holes drilled in connection with exploration for oil encounter waters saturated with soluble salts. The composition of one such water is compared in Table 12.8 with that of highly saline surface waters (Great Salt Lake, Dead Sea). Several rather important differences are seen. For example, the HCO_3^- content of the brine is about 10 percent of the total anion content. Sulfate ion, an important component of surface water, is essentially absent in the deep brines.

TABLE 12.8

Comparison of the Composition of Two Hypersaline Lakes and an Oil-field Brine with That of Seawater (All Concentrations Are Milligrams per Liter)

Component	Seawater[†]	Great Salt Lake[‡]	Dead Sea[§]	Oil-field Brine[¶]
Na^+	10,600	67,500	35,000	9,400
K^+	380	3,380	7,560	124
Ca^{++}	400	330	15,800	177
Mg^{++}	1,270	5,600	42,000	127
Cl^-	19,000	113,000	208,000	14,400
Br^-	65	. . . .	5,900	99
SO_4^{--}	2,650	13,600	540	1
HCO_3^-	140	180	240	1,500

[†] H. U. Sverdrup, M. W. Johnson, and R. H. Fleming, "The Oceans," Prentice-Hall, Inc., Englewood Cliffs, N.J., 1957.
[‡] R. Rankama and Th. G. Sahama, "Geochemistry," The University of Chicago Press, Chicago, 1955.
[§] Y. K. Bentor, Some Geochemical Aspects of the Dead Sea and the Question of Its Age, *Geochim. Cosmochim. Acta*, 25: 239 (1961).
[¶] D. E. White, Magmatic, Connate and Metamorphic Waters, *Geol. Soc. Am. Bull.* 68, p. 1659, 1957.

PROBLEMS

12.1 Two dilute solutions of NaOH were prepared by dissolving 0.04 g of solid NaOH in water. For solution A, distilled water was used. For solution B, distilled water was boiled just prior to use in making the NaOH solution. Each solution was used to titrate 25-ml aliquots of HCl (0.001 m). It was found that a larger quantity of solution A was needed. Explain the result.

12.2 In concentrated HCl solution Fe^{++} and Fe^{3+} form strong chloride complexes. Would this phenomenon affect the oxidation state of iron in a solution open to the atmosphere? Explain your answer.

*12.3 Using free energy data for the minerals calcite and dolomite and the activities of their constituent ions in seawater, which should be the more stable phase in deep-sea sediments? What would the Mg/Ca ratio in seawater have to be for the two phases to be equally stable?

12.4 If a saline lake is precipitating $CaCO_3$ and $MgSiO_3$ and is just at saturation for both minerals, show that the Mg/Ca ratio is fixed by the dissolved silica content of the lake. Assume that all Si in the lake is in the neutral form H_4SiO_4 and that the p_{CO_2} for the lake water equals that in the atmosphere.

*12.5 Given the following information, estimate the degree of saturation of $BaSO_4$ in the deep sea:

$$[\Sigma SO_4] = 0.27 \text{ g/liter}$$
$$[\Sigma Ba] = 25 \ \mu\text{g/liter}$$
$$K_{BaSO_4} = 10^{-2.3} \quad \text{for major-ion complexing}$$

Other necessary data may be found in Garrels and Christ.

12.6 As long as no other weak bases are present in significant amounts, the total carbonate-ion concentration in any brine can be determined as follows: The ΣCO_2 and p_{CO_2} of the brine are measured. A known amount of KOH is added and the p_{CO_2} is redetermined. Show that the ΣCO_3^{--} content is given from the relationship

$$\frac{p'_{CO_2}}{p_{CO_2}} = \left(\frac{[\Sigma CO_2] - [\Sigma CO_3^{--}] - \Delta}{[\Sigma CO_2] - [\Sigma CO_3^{--}]} \right)^2 \frac{[\Sigma CO_3^{--}]}{[\Sigma CO_3^{--} + \Delta]}$$

where p_{CO_2} and p'_{CO_2} are the CO_2 partial pressures measured before and after the base addition, and Δ is the amount of base added. The assumptions are made that the amount of CO_2 in the form of CO_2 gas is negligible and that a negligible amount of the added hydroxyl ion goes into raising the pH of the solution. Since the addition of a small amount of KOH leaves the matrix of the solution unchanged, the activity coefficients and degree of major-ion complexing do not significantly change.

REFERENCES

Garrels, R. M., and C. L. Christ: "Solutions, Minerals, and Equilibria," Harper & Row, Publishers, Incorporated, New York, 1965.
—— and M. E. Thompson: *Am. J. Sci.*, **260**: 57 (1962).
Helgeson, H. C.: *Am. J. Sci.*, **267**: 729 (1969).
Klotz, I. M.: "Chemical Thermodynamics," Prentice-Hall, Inc., Englewood Cliffs, N.J., 1950.

SUPPLEMENTARY READING

Culberson, C., D. R. Kester, and R. M. Pytkowicz: High Pressure Dissociation of Carbonic and Boric Acids in Sea Water, *Science*, **157**: 59 (1967).
Helgeson, H. C.: Solution Chemistry and Metamorphism, in P. H. Abelson (ed.), "Researches in Geochemistry," p. 362, John Wiley & Sons, Inc., New York, 1967.
——: Evaluation of Irreversible Reactions in Geochemical Processes Involving Minerals and Aqueous Solutions. I, Thermodynamic Relations, *Geochim. Cosmochim. Acta*, **32**: 853 (1968).
Hostetlier, P. B., and C. L. Christ: Studies in the System MgO-SiO_2-CO_2-H_2O(I): The Activity Product Constant of Chrysotile, *Geochim. Cosmochim. Acta*, **32**: 485 (1968).
Lerman, A.: Model of Chemical Evolution of a Chloride Lake—The Dead Sea, *Geochim. Cosmochim. Acta*, **31**: 2309 (1967).
Li, Y.-H., T. Takahashi, and W. S. Broecker: The Degree of Saturation of $CaCO_3$ in the Oceans, *J. Geophys. Res.*, **74**: 5507 (1969).
Paces, T.: Chemical Equilibria and Zoning of Subsurface Water from Jachymov Ore Deposits, Czechoslovakia, *Geochim. Cosmochim. Acta*, **33**: 591 (1969).
Sillen, L. G.: The Ocean as a Chemical System, *Science*, **156**: 1189 (1967).
Takahashi, T., W. Broecker, Y.-H. Li, and D. Thurber: Chemical and Isotopic Balances for a Meromictic Lake, *Limnol. Oceanogr.*, **13**: 272 (1968).

appendix I Thermodynamic Derivations

A. THERMODYNAMIC EQUATIONS OF STATE: $dE = T\,dS - P\,dV$

The first law of thermodynamics states that energy can be neither created nor lost. Consequently, the change in energy of a system is equal to the work done on the system by outside forces plus the heat added to the system:

$$dE = \delta q + \delta w$$

For systems capable of only pressure-volume work, the work done on the system can be expressed as

$$\delta w = -P\,dV$$

If the volume of the system decreases at constant pressure and temperature, work has been done on the system and δw is positive.

The second law of thermodynamics can be stated in one of its classical forms as

$$dS = \frac{\delta q_{\text{rev}}}{T}$$

The differential change in entropy of the system is equal to the heat added to the system under reversible conditions at constant temperature divided by the temperature of addition.

Combining the classical statements of the first and second laws gives

$$dE = T\,dS - P\,dV$$

for a system capable only of pressure-volume work. For systems where other forms of work are possible, such as by release of strain energy, other work terms must be added.

B. DIFFERENTIALS OF OTHER THERMODYNAMIC FUNCTIONS

1. Enthalpy

$$H \equiv E + PV$$

$$dH = dE + P\,dV + V\,dP$$

But

$$dE = T\,dS - P\,dV$$

and so

$$dH = T\,dS - P\,dV + P\,dV + V\,dP$$

$$dH = T\,dS + V\,dP$$

2. Helmholtz free energy

$$A \equiv E - TS$$

$$dA = dE - T\,dS - S\,dT$$

$$dA = T\,dS - P\,dV - T\,dS - S\,dT$$

$$dA = -P\,dV - S\,dT$$

3. Gibbs free energy

$$G \equiv H - TS$$

$$dG = dH - T\,dS - S\,dT$$

$$dG = T\,dS + V\,dP - T\,dS - S\,dT$$

$$dG = V\,dP - S\,dT$$

C. EQUALITY OF PARTIAL DERIVATIVES

A theorem of calculus concerning functions of more than one variable states that for a function $Z(X,Y)$, where Z is the function and X and Y

are the independent variables, the crossed partial derivatives are equal under certain conditions.   The total differential of Z is

$$dZ = \left(\frac{\partial Z}{\partial X}\right)_Y dX + \left(\frac{\partial Z}{\partial Y}\right)_X dY$$

If the total differential of Z is an exact differential, the theorem of crossed partials states that

$$\left[\frac{\partial(\partial Z/\partial X)_Y}{\partial Y}\right]_X = \left[\frac{\partial(\partial Z/\partial Y)_X}{\partial X}\right]_Y$$

Fortunately dE, dH, dA, and dG are exact differentials.   Taking the crossed partial derivatives of these functions gives

$$\left(\frac{\partial T}{\partial V}\right)_S = -\left(\frac{\partial P}{\partial S}\right)_V \qquad \text{from } dE$$

$$\left(\frac{\partial T}{\partial P}\right)_S = \left(\frac{\partial V}{\partial S}\right)_P \qquad \text{from } dH$$

$$\left(\frac{\partial P}{\partial T}\right)_V = \left(\frac{\partial S}{\partial V}\right)_T \qquad \text{from } dA$$

$$\left(\frac{\partial V}{\partial T}\right)_P = -\left(\frac{\partial S}{\partial P}\right)_T \qquad \text{from } dG$$

This set of equalities is called the Maxwell relations.

D. CHANGE OF VARIABLES

The total differential of X is

$$dX = \left(\frac{\partial X}{\partial Y}\right)_Z dY + \left(\frac{\partial X}{\partial Z}\right)_Y dZ$$

If $dX = 0$

$$\left[-\left(\frac{\partial X}{\partial Y}\right)_Z dY = \left(\frac{\partial X}{\partial Z}\right)_Y dZ\right]_{dX=0}$$

or

$$\left(\frac{\partial X}{\partial Y}\right)_Z = -\left(\frac{\partial X}{\partial Z}\right)_Y \left(\frac{\partial Z}{\partial Y}\right)_X$$

E. THE RELATION $(\partial S/\partial P)_T = -\alpha V$

From the Maxwell relations,

$$\left(\frac{\partial S}{\partial P}\right)_T = -\left(\frac{\partial V}{\partial T}\right)_P$$

By definition

$$\alpha \equiv \frac{1}{V}\left(\frac{\partial V}{\partial T}\right)_P$$

and

$$\alpha V = \left(\frac{\partial V}{\partial T}\right)_P$$

Thus

$$\left(\frac{\partial S}{\partial P}\right)_T = -\alpha V$$

F. TEMPERATURE DERIVATIVES OF ENTROPY

The principal heat capacities are defined as

$$C_V \equiv \left(\frac{\delta q}{dT}\right)_V$$

$$C_P \equiv \left(\frac{\delta q}{dT}\right)_P$$

But

$$\delta q_{\text{rev}} = T\, dS$$

and

$$\frac{\delta q}{dT} = T\left(\frac{\partial S}{\partial T}\right)$$

Thus

$$C_V = T\left(\frac{\partial S}{\partial T}\right)_V$$

and

$$C_P = T\left(\frac{\partial S}{\partial T}\right)_P$$

or

$$\left(\frac{\partial S}{\partial T}\right)_V = \frac{C_V}{T}$$

$$\left(\frac{\partial S}{\partial T}\right)_P = \frac{C_P}{T}$$

G. TOTAL DERIVATIVE OF ENTROPY FOR T AND V AS VARIABLES

$$dS = \left(\frac{\partial S}{\partial T}\right)_V dT + \left(\frac{\partial S}{\partial V}\right)_T dV$$

From above

$$\left(\frac{\partial S}{\partial T}\right)_V = \frac{C_V}{T}$$

From the Maxwell relations,

$$\left(\frac{\partial S}{\partial V}\right)_T = \left(\frac{\partial P}{\partial T}\right)_V$$

Changing variables,

$$\left(\frac{\partial P}{\partial T}\right)_V = - \left(\frac{\partial P}{\partial V}\right)_T \left(\frac{\partial V}{\partial T}\right)_P$$

But

$$\alpha \equiv \frac{1}{V} \left(\frac{\partial V}{\partial T}\right)_P$$

$$\beta \equiv - \frac{1}{V} \left(\frac{\partial V}{\partial P}\right)_T$$

and so

$$\left(\frac{\partial P}{\partial T}\right)_V = \frac{\alpha}{\beta}$$

and

$$dS = \frac{C_V}{T} dT + \frac{\alpha}{\beta} dV$$

H. TOTAL DERIVATIVE OF ENTROPY FOR T AND P AS VARIABLES

$$dS = \left(\frac{\partial S}{\partial T}\right)_P dT + \left(\frac{\partial S}{\partial P}\right)_T dP$$

From above

$$\left(\frac{\partial S}{\partial T}\right)_P = \frac{C_P}{T}$$

From Sec. F

$$\left(\frac{\partial S}{\partial P}\right)_T = -\alpha V$$

and so

$$dS = \frac{C_P}{T} dT - \alpha V \, dP$$

I. ADIABATIC GRADIENT, $(\partial T/\partial P)_S$

Changing variables

$$\left(\frac{\partial T}{\partial P}\right)_S = -\left(\frac{\partial T}{\partial S}\right)_P \left(\frac{\partial S}{\partial P}\right)_T$$

From above

$$\left(\frac{\partial T}{\partial S}\right)_P = \frac{1}{(\partial S/\partial T)_P} = \frac{T}{C_P}$$

and

$$\left(\frac{\partial S}{\partial P}\right)_T = -\alpha V$$

and so

$$\left(\frac{\partial T}{\partial P}\right)_S = \frac{TV\alpha}{C_P}$$

J. TOTAL DIFFERENTIAL OF ENERGY IN TERMS OF T AND V

$$dE = T\,dS - P\,dV \qquad \text{from Sec. A}$$

$$T\,dS = C_V\,dT + \frac{T\alpha}{\beta}\,dV \qquad \text{from Sec. G}$$

Thus

$$dE = C_V\,dT + \left(\frac{T\alpha}{\beta} - P\right)dV$$

and

$$\left(\frac{\partial E}{\partial T}\right)_V = C_V$$

$$\left(\frac{\partial E}{\partial V}\right)_T = \frac{T\alpha}{\beta} - P$$

K. TOTAL DIFFERENTIAL OF ENTHALPY IN TERMS OF P AND T

$$dH = T\,dS + V\,dP \qquad \text{from Sec. B}$$

$$T\,dS = C_P\,dT - T\alpha V\,dP \qquad \text{from Sec. H}$$

Thus

$$dH = C_P\,dT + V(1 - \alpha T)\,dP$$

and

$$\left(\frac{\partial H}{\partial T}\right)_P = C_P$$

$$\left(\frac{\partial H}{\partial P}\right)_T = V(1 - \alpha T)$$

L. DIFFERENCE BETWEEN PRINCIPAL HEAT CAPACITIES

From above

$$T \, dS = C_P \, dT - T\alpha V \, dP$$

and

$$T \, dS = C_V \, dT - \left(\frac{\alpha T}{\beta} - P\right) dV$$

Combining

$$C_P \, dT - T\alpha V \, dP = C_V \, dT - \left(\frac{\alpha T}{\beta} - P\right) dV$$

$$(C_P - C_V) \, dT = T\alpha V \, dP - \left(\frac{\alpha T}{\beta} - P\right) dV$$

$$dT = \frac{T\alpha V \, dP}{C_P - C_V} - \frac{(\alpha T/\beta - P) \, dV}{C_P - C_V}$$

dT can be expressed as

$$dT = \left(\frac{\partial T}{\partial V}\right)_P dV + \left(\frac{\partial T}{\partial P}\right)_V dP$$

Equating the dP terms,

$$\frac{T\alpha V \, dP}{C_P - C_V} = \left(\frac{\partial T}{\partial P}\right)_V dP$$

But

$$\left(\frac{\partial T}{\partial P}\right)_V = \frac{\beta}{\alpha} \qquad \text{from Sec. G}$$

and so

$$C_P - C_V = \frac{TV\alpha^2}{\beta}$$

M. THE VALUE OF $(\partial E/\partial T)_P$

From

$$H = E + PV$$

$$\left(\frac{\partial H}{\partial T}\right)_P = \left(\frac{\partial E}{\partial T}\right)_P + P\left(\frac{\partial V}{\partial T}\right)_P$$

$$\left(\frac{\partial E}{\partial T}\right)_P = C_P - P\alpha V$$

But

$$C_P = C_V + \frac{TV\alpha^2}{\beta}$$

and so

$$\left(\frac{\partial E}{\partial T}\right)_P = C_V + \frac{TV\alpha^2}{\beta} - P\alpha V$$

N. TEMPERATURE DEPENDENCE OF THE DISTRIBUTION COEFFICIENT

$$\left(\frac{\partial \ln K_D}{\partial T}\right)_P = \frac{\Delta H}{RT^2}$$

Consider component A in phases α and β:

$$\mu_A{}^\alpha = \mu_A^{*\alpha} + RT \ln \gamma_A{}^\alpha x_A{}^\alpha$$

$$\mu_A{}^\beta = \mu_A^{*\beta} + RT \ln \gamma_A{}^\beta x_A{}^\beta$$

where μ_A^* is the chemical potential of pure component A at the temperature and pressure of interest. At equilibrium

$$\mu_A{}^\alpha = \mu_A{}^\beta$$

and

$$\mu_A^{*\alpha} + RT \ln \gamma_A{}^\alpha x_A{}^\alpha = \mu_A^{*\beta} + RT \ln \gamma_A{}^\beta x_A{}^\beta$$

The distribution coefficient K_D is defined as

$$K_D = \frac{\gamma_A{}^\beta x_A{}^\beta}{\gamma_A{}^\alpha x_A{}^\alpha}$$

Thus

$$\mu_A^{*\alpha} - \mu_A^{*\beta} = RT \ln K_D$$

$$\frac{\partial(\mu_A^{*\alpha}/T)}{\partial T} = -\frac{\bar{H}_A{}^\alpha}{T^2} \qquad \text{for } \gamma = 1$$

where $\bar{H}_A{}^\alpha$ is the partial molar enthalpy of component A in phase α.
In the limit of ideal solutions

$$\frac{\partial \ln K_D}{\partial T} = -\frac{\bar{H}_A{}^\alpha}{RT^2} + \frac{\bar{H}_A{}^\beta}{RT^2}$$

Defining the difference in partial molar enthalpies $\bar{H}_A{}^\beta - \bar{H}_A{}^\alpha$ as ΔH

$$\frac{\partial \ln K_D}{\partial T} = \frac{\Delta H}{RT^2}$$

O. PRESSURE DEPENDENCE OF THE EQUILIBRIUM CONSTANT

$$R \ln K = -\Sigma \nu_i \frac{\mu_i^*}{T}$$

$$\frac{R \partial \ln K}{\partial P} = -\Sigma \nu_i \left(\frac{\partial \mu_i^* / T}{\partial P} \right)_T$$

The total differential of free energy is

$$dG = -S \, dT + V \, dP + \Sigma \mu_i \, dn_i$$

$$dG = -S \, dT + V \, dP + \Sigma \mu_i^* \, dn_i + \Sigma RT \ln \gamma_i x_i \, dn_i$$

Taking crossed partial derivatives,

$$\left(\frac{\partial \mu_i^*}{\partial P} \right)_{n_i, T} = \left(\frac{\partial V}{\partial n_i} \right)_{P,T} \equiv \bar{V}_i$$

Thus

$$R \frac{\partial \ln K}{\partial P} = -\Sigma \nu_i \frac{\bar{V}_i}{T}$$

But

$$\Sigma \nu_i \bar{V}_i = \Delta V° \qquad \text{the volume change for the reaction in the limit of ideal behavior}$$

and

$$\left(\frac{\partial \ln K}{\partial P} \right)_T = -\frac{\Delta V°}{RT}$$

appendix II List of Symbols

a	Acceleration
a	Activity
a	Concentration
a	Distance
a	Orthogonal dimension
a	Frequently used for constant in equations
A	Alkalinity
A	Amount of species
A	Area
A	Arrhenius reaction rate
A	Frequently used as constant in equations
A	Helmholtz free energy
A	Product of anion and cation charges
A	Total wave function
A	Virial coefficient
b	Frequently used for constant in equations
B	Amount of species
B	Bulk modulus
B	Frequently used for constant in equations

B	Virial coefficient
c	Speed of light
C	Centigrade
C	Concentration
C	Constant
C	Heat capacity
C	Number of components in phase rule
C	Vibration rate
C	Virial coefficient
d	Atomic or molecular distance
D	Constant related to size of ion in activity-coefficient equation
D	Diffusion constant
D	Virial coefficient
e	Electronic charge
e	Exponential function
exp	Exponential function
E	Energy
E_B	Barrier energy for reaction
E_D	Barrier energy for diffusion
f	Efficiency factor for heat-work exchange
f	Fraction of free volume
f	Fugacity
f	Number of variables required to uniquely define system
$f_{C.P.}$	Flux through center plane
$f_{E_B \to \infty}$	Fraction of molecules with $E > E_B$
f_i	Fraction of molecules in energy level i
F	Flux
F	Force
g	Gravitational acceleration
G	Gibbs free energy
$\bar{G}$	Molar free energy
h	Elevation
h	Planck's constant
H	Enthalpy
$\bar{H}$	Molar enthalpy
i	Energy level
i	Ion
i	Substance i—chemical species
j	Positive integer
j	Rotational quantum number
k	Boltzmann's constant
k	Force constant for vibration
k	Henry's law constant

k	Proportionality constant in Arrhenius rate law
K_D	Distribution coefficient
K	Kelvin, degrees absolute
K_f	Equilibrium constant calculated with fugacities
K_p	Equilibrium constant calculated with partial pressures
$K_{\text{S.P.}}$	Solubility-product constant
l	Length
m	Constant in diffusion equation
m	Exponent in rate law
m	Mass of diffusing substance
m	Molal concentration
m	Molecular mass
M	Amount of material
n	Any integer
n	Constant in diffusion law
n	Exponent in ionic force law
n	Exponent in rate law
n	Number of atoms
n	Quantum number
n_i	Ionic charge on species i
n_i	Number of molecules in substance i
N	Number of atoms
N	Total number of moles in system
N_0	Avogadro's number
N_{tot}	Number of defects in a solid
p	Momentum
p	Partial pressure
p	Probability
P	Number of phases in phase rule
P	Order of reaction
P	Pressure
$P_{\text{equil}}^{X^\circ C}$	Pressure of a phase change at given temperature
q	Single-particle partition function
q_C	Heat transfer from cold reservoir
q_H	Heat transfer from hot reservoir
r	Bond length
r	Distance of separation of masses
r	Molecular radius
R	Universal gas constant
R'	Rate
$\bar{S}$	Degree of saturation
S	Entropy
$\bar{S}$	Molar entropy

$\bar{S}_i$	Partial molal entropy of substance i
t	Time
$t_{\frac{1}{2}}$	Half-life of radioactive species
T	Temperature
T_B	Boyle temperature
u	Potential energy
v	Velocity
$\bar{v}$	Average velocity
V	Volume
$\bar{V}$	Molar volume
V_B	Molar volume of substance B
V_0	Volume at absolute zero
$\bar{V}_i$	Partial molal volume of substance i
$\bar{V}_i^{\infty}$	Partial molal volume of substance i at infinite dilution
w	Work
W	Complexions
W	Free energy constant
W_X	Constant in excess function for thermodynamic variable X
x	Distance of separation between ions
x	Distance parameter
x	Mole fraction
x_{equil}	Equilibrium distance of separation between ions
X	Function of free volume
$\bar{X}_i$	Partial molal variable (X) for substance i
Z	Compressibility factor
Z	Number of collisions per second per cubic centimeter in a gas
Z_A	Anion charge
Z_C	Cation charge

SUBSCRIPTS

A	Component, compound, element, molecule, phase, or system
B	Component, compound, element, molecule, phase, or system
c	Critical
C	Cold
cpx	Clinopyroxene
crist	Cristobalite
cryst	Crystal
D	Debye
E	Einstein
equil	Equilibrium
ext	External
f	Fusion
G	Glacier

H	Heavy
H	Hole
H	Hot
int	Internal
j	Other substances
L	Light
liq	Liquid
lith	Lithostatic
mech. mix.	Mechanical mixture
p	Ratio of partial pressures
P	Constant pressure
P.B.	Phase boundary
perm	Permanent
prod	Product
qtz	Quartz
r	Reduced
reac	Reaction (or reactant)
rev	Reversible
rot	Rotation
sol	Solid
soln	Solution
S.P.	Solubility product
std	Standard
sur	Surroundings
sys	System
T	Isothermal
TI	Translation and indistinguishability
Tx	Translation for x direction
tot	Total
trans	Translation
univ	Universe
V	Constant volume
vib	Vibration
x	Constant x
xs	Excess
0	Ground state

SUPERSCRIPTS

ab	Albite
an	Anorthite
fl	Fluid
neph	Nepheline
xl	Crystal

GREEK

α	Separation constant for isotope reactions
α	Thermal-expansion coefficient
β	Isothermal compressibility
γ	Activity coefficient for liquids and solids
γ	C_P/C_V
γ^*	Apparent activity coefficient for ions in aqueous solution
Γ	Activity coefficient for gases
δ	Change in isotopic composition
Δ	Xproducts $-$ Xreactants
ϵ	Energy
θ	Characteristic temperature
θ_D	Debye temperature
θ_E	Einstein temperature
λ	Decay constant
λ	Wavelength
μ	Chemical potential
μ	Ionic strength
μ	Reduced mass
ν	Frequency
ν_i	Coefficient in chemical reactions
Π	Product
ρ	Density
φ	Potential
ψ	Space part of Schrödinger equation
ψ	Wave function

Index